# Titles in This Series

Volume

# Titles in This Series

**Volume**

## Titles in This Series

**Volume**

Mladen Luksic, Clyde Martin, and William Shadwick, Editors

69 **Methods and applications of mathematical logic,** Walter A. Carnielli and Luiz Paulo de Alcantara, Editors

70 **Index theory of elliptic operators, foliations, and operator algebras,** Jerome Kaminker, Kenneth C. Millett, and Claude Schochet, Editors

71 **Mathematics and general relativity,** James A. Isenberg, Editor

72 **Fixed point theory and its applications,** R. F. Brown, Editor

73 **Geometry of random motion,** Rick Durrett and Mark A. Pinsky, Editors

74 **Geometry of group representations,** William M. Goldman and Andy R. Magid, Editors

75 **The finite calculus associated with Bessel functions,** Frank M. Cholewinski

76 **The structure of finite algebras,** David C. Hobby and Ralph Mckenzie

77 **Number theory and its applications in China,** Wang Yuan, Yang Chung-chun, and Pan Chengbiao, Editors

78 **Braids,** Joan S. Birman and Anatoly Libgober, Editors

79 **Regular differential forms,** Ernst Kunz and Rolf Waldi

80 **Statistical inference from stochastic processes,** N. U. Prabhu, Editor

81 **Hamiltonian dynamical systems,** Kenneth R. Meyer and Donald G. Saari, Editors

82 **Classical groups and related topics,** Alexander J. Hahn, Donald G. James, and Zhe-xian Wan, Editors

83 **Algebraic K-theory and algebraic number theory,** Michael R. Stein and R. Keith Dennis, Editors

84 **Partition problems in topology,** Stevo Todorcevic

85 **Banach space theory,** Bor-Luh Lin, Editor

86 **Representation theory and number theory in connection with the local Langlands conjecture,** J. Ritter, Editor

87 **Abelian group theory,** Laszlo Fuchs, Rüdiger Göbel, and Phillip Schultz, Editors

88 **Invariant theory,** R. Fossum, W. Haboush, M. Hochster, and V. Lakshmibai, Editors

89 **Graphs and algorithms,** R. Bruce Richter, Editor

# Graphs and Algorithms

CONTEMPORARY MATHEMATICS

Volume 89

# Graphs and Algorithms

Proceedings of the AMS-IMS-SIAM Joint Summer Research Conference held June 28–July 4, 1987 with support from the National Science Foundation

R. Bruce Richter, Editor

AMERICAN MATHEMATICAL SOCIETY
Providence · Rhode Island

The AMS-IMS-SIAM Joint Summer Research Conference in the Mathematical Sciences on Graphs and Algorithms was held at the University of Colorado, Boulder, Colorado on June 28–July 4, 1987 with support from the National Science Foundation, Grant DMS-8613199.

1980 *Mathematics Subject Classification* (1985 *Revision*). Primary 05C, 68M, 68P, 68Q.

---

**Library of Congress Cataloging-in-Publication Data**

AMS-IMS-SIAM Joint Summer Research Conference in the Mathematical Sciences on Graphs and Algorithms (1987: University of Colorado)

Graphs and algorithms: proceedings of the AMS-IMS-SIAM joint summer research conference held June 28–July 4, 1987 with support from the National Science Foundation /R. Bruce Richter, editor.

p. cm.–(Contemporary mathematics, ISSN 0271-4132; v. 89)

"The AMS-IMS-SIAM Joint Summer Research Conference in the Mathematical Sciences on Graphs and Algorithms was held at the University of Colorado, Boulder, Colorado"–T.p. verso.

Includes bibliographies.

ISBN 0-8218-5095-4 (alk. paper)

1. Graph theory–Congresses. 2. Algorithms–Congresses. I. Richter, R. Bruce. II. American Mathematical Society. III. Institute of Mathematical Statistics. IV. Society for Industrial and Applied Mathematics. V. Title. VI. Series: Contemporary mathematics (American Mathematical Society); v. 89.

QA166.A47 1987 — 89-216

511′.5–dc19 — CIP


---

# CONTENTS

## FOREWORD

The **GRAPHS AND ALGORITHMS** week of the Joint Summer Research Conference in the Mathematical Sciences took place in Boulder, Colorado, June 28 to July 4, 1987. The co-chairs were Joe Buhler and Phyllis Chinn.

The purpose of the week was to foster communication between computer scientists and mathematicians. Recent work in graph theory and related algorithms has relied on increasingly sophisticated mathematics. Wagner's Conjecture, self-adjusting data structures, graph isomorphism and various embedding and labelling problems in VLSI are examples. Thus, it is clear that researchers in both computer science and mathematics need to be aware of what the other is doing.

The organizers were pleasantly surprised at the depth and diversity of the work being done. The wide range of talks demonstrated the vitality of the activity in both fields. The common graph-theoretic and algorithmic themes insured sufficient cross-interest so many people profited from discussions with others whose work had led in different directions. We hope the papers included here capture some of the diversity and excitement of the conference.

The organizing committee would like to thank all those who attended for their participation. We would also like to thank the AMS-IMS-SIAM Committee and the National Science Foundation for their support. We especially appreciate the hard work of Carole Kohanski of the AMS. The University of Colorado at Boulder was an excellent host institution.

The Organizing Committee

## ORGANIZING COMMITTEE

**co-chairs:** Joe P. Buhler (Reed College)
Phyllis Z. Chinn (Humboldt State University)

**committee:** Jean Larson (University of Florida)
Eugene Luks (University of Oregon)
Bruce Richter (U.S. Naval Academy)
Bruce Rothschild (University of California at Los Angeles)
Robert Tarjan (AT&T Bell Laboratories)

## LIST OF PARTICIPANTS

Curtiss Barefoot
New Mexico Institute of Mining and Technology
Socorro, New Mexico

Larry I. Basenspiler
Northern Illinois University
DeKalb, Illinois

Fredricka T. Bennett
Skidmore College
Saratoga Springs, New York

David R. Berman
University of North Carolina
Wilmington, North Carolina

Ken D. Blaha
University of Oregon
Eugene, Oregon

Gary M. Brenner
P.O. Box 7114
Boulder, Colorado

Thomas C. Brown
Simon Fraser University
Burnaby, British Columbia

Joe P. Buhler
M.S.R.I. Reed College
Portland, Oregon

Bruce A. Chalmers
Mt. Allison University
Sackville, New Brunswick

Eldon K.W. Chan
San Jose State University
San Jose, California

Phyllis Z. Chinn
Humboldt State University
Arcata, California

Fan R.K. Chung
Bell Communications Research
Morristown, New Jersey

Karen L. Collins
Wesleyan University
Middletown, Connecticut

Alice M. Dean
Skidmore College
Saratoga Springs, New York

Jane W. Di Paola
3580 County Rd. 215
Cheyenne, Wyoming

Clifton E. Ealy, Jr.
Northern Michigan University
Marquette, Michigan

Vance Faber
Los Alamos National Laboratory
Los Alamos, New Mexico

Arthur M. Farley
University of Oregon
Eugene, Oregon

Joan Feigenbaum
600 Mountain Ave.
Murray Hill, New Jersey

Mike R. Fellows
University of Idaho
Moscow, Idaho

Michael D. Fried
University of Florida
Gainesville, Florida

Mark K. Goldberg
Rensselaer Polytechnic Institue
Troy, New York

Steven B. Grantham
Boise State University
Boise, Idaho

Alvin C. Green
State University College
at Buffalo
Buffalo, New York

Charles M. Grinstead
Swarthmore College
Swarthmore, Pennsylvania

Robert L. Grossman
University of California
Berkeley, California

Stanley J. Gurak
University of San Diego
San Diego, California

Eleanor O. Hare
Clemson Unviersity
Clemson, South Carolina

William R. Hare
Clemson University
Clemson, South Carolina

Sandra M. Hedetniemi
Clemson University
Clemson, South Carolina

Steven T. Hedetniemi
Clemson University
Clemson, South Carolina

Kim A.S. Hefner
University of Colorado
Denver, Colorado

Arthur M. Hobbs
Texas A&M University
College Station, Texas

Fred Hoffman
Florida Atlantic University
Boca Raton, Florida

Joan P. Hutchinson
Smith College
Northampton, Massachusetts

Brad W. Jackson
San Jose State University
San Jose, California

Kathryn L. Jones
University of Colorado
Denver, Colorado

Philip N. Klein
Massachusetts Institute
of Technology
Cambridge, Massachusetts

Yueh-Er Kuo
University of Tennessee
Knoxville, Tennessee

Jean A. Larson
University of Florida
Gainesville, Florida

Renu Laskar
Clemson University
Clemson, South Carolina

Sin-Min Lee
San Jose State University
San Jose, California

Eugene M. Luks
University of Oregon
Eugene, Oregon

J. Richard Lundgren
University of Colorado
Denver, Colorado

Mary A. Maher
New Mexico State University
Las Cruces, New Mexico

Peter D. Mark
University of Oregon
Eugene, Oregon

Ursula Martin
RHBNC, University of London
Egham, Surrey

Robert A. Melter
Long Island University
Southampton, New York

Zevi Miller
Miami University
Oxford, Ohio

John A. Mitchem
San Jose State University
San Jose, California

Eric Regener
Concordia University
Montreal, Quebec

R. Bruce Richter
U.S. Naval Academy
Annapolis, Maryland

Arnold L. Rosenberg
University of Massachusetts
Amherst, Massachusetts

Michael E. Saks
Rutgers University
New Brunswick, New Jersey

Peter J. Slater
Clemson University
Clemson, South Carolina

Daniel D. Sleator
Carnegie Mellon University
Pittsburgh, Pennsylvania

Joseph B. Stephen
Northern Illinois University
DeKalb, Illinois

Rebekka R. Struik
University of Colorado
Boulder, Colorado

Jayme L. Szwarcfiter
Universidad Federal do Rio
de Janeiro
Rio de Janeiro, Brazil

William T. Trotter
Arizona State University
Tempe, Arizona

Miroslaw Truszczynski
University of Kentucky
Lexington, Kentucky

Donald W. Vandejagt
Grand Valley State College
Allendale, Michigan

Andrew Vince
University of Florida
Gainesville, Florida

Walter D. Wallis
Southern Illinois University
Carbondale, Illinois

Michael Werman
Brown University
Providence, Rhode Island

Indra Wui
San Jose State University
San Jose, California

Contemporary Mathematics
Volume **89**, 1989

# THE ROBERTSON-SEYMOUR THEOREMS: A SURVEY OF APPLICATIONS

Michael R. Fellows

**Abstract** Applications of the Robertson-Seymour theorems to a variety of problems in concrete computational complexity are surveyed. These results suggest a broad program of research in computational complexity based on partial orders of combinatorial objects defined by sequences of local operations. It is shown that the 2–induced connecting paths problem, relevant to the problem of order-testing in the strong minor order, is $NP$-complete. Approaches to practical algorithms achieving the Robertson-Seymour complexity bounds are described.

## 1. Introduction

It is almost 100 years since Hilbert's now famous, and then controversial, nonconstructive solution to Gordon's Problem of Invariants. Hilbert proved the following finite basis theorem.

**Theorem**

(Hilbert) Any set of forms in a finite number of variables over $Q(\alpha)$ has a finite basis. □

According to a recent historical account by Beeson [Be] Hilbert corresponded with Cayley for over a month to convince him that he had actually solved the problem. The nonconstructive nature of Hilbert's argument elicited from Gordon the famous reaction [Re],

"This is not Mathematics, this is Theology!"

Hilbert's proof was controversial because until that time mathematical argument had been almost entirely computational in nature (and therefore constructive). He had proved that a finite basis *must exist*, without giving any information on how to *compute* it, and this ran against the grain of mathematical intuition in that day. The controversy led


0271-4132/89 $1.00 + $.25 per page

historically to several things. Partly because of it, Hilbert formulated his famous Program, shattered by Gödel, and in some sense resurrected by the $P=NP$ question. According to Beeson, the proof and the reaction that followed helped to crystalize two schools of mathematical philosophy that developed, actually, in tandem — the "classical" (but, really, modern) set-theoretic outlook, the majority view that we now take for granted, and the minority view of the constructivists. Peace has been somewhat restored between these two schools as it has become clear that both kinds are fun *to do* [BR].

Computer Science, for the 30 years or so of its existence, has been a modern haven for naive constructivism, such as was the prevailing outlook in Mathematics one hundred years ago. It has developed in some measure independently of the other great branches of "self-conscious" mathematical endeavor, Logic and Constructivism, primarily because of the tremendous roots in applications that Computer Science possesses.

The recent solution of Wagner's Conjecture in the affirmative by Neil Robertson and Paul Seymour presents us with one of the deepest and most beautiful theorems in all of Mathematics. Their combinatorial finite basis theorem is stated as follows.

**Theorem** (Robertson-Seymour) Any set of finite graphs has a finite set of minimal elements in the minor ordering.

The nonconstructive nature of their proof and their parallel results on the computational complexity of order testing in the minor order allow us to relive, in some measure, that scandal of a century ago.

Echoing Gordon, a prominent computer scientist of our day has reacted to these developments,

"This is not Computer Science, this is a Mathematical Curiosity!"

The scandal is that we are now provided with enormously general and powerful tools for proving small-degree polynomial-time complexity for a wide variety of problems of combinatorial optimization — nonconstructively. Proofs of polynomial-time complexity employing the Robertson-Seymour theorems are actually nonconstructive on two distinct levels.

In the first place, an argument employing these results establishes only that an algorithm exists—we are not told what the algorithm is, nor are we provided with any effective means of finding it. In the second place, even if the algorithm were known, it would correctly decide, but not on the basis of finding, or failing to find, natural evidence. For example, we might be told (efficiently) by such a decision algorithm that a permutation of the vertex set with certain properties exists, without being given any hint how such a permutation might be (efficiently) found. Both of these forms of nonconstructive existence assertion, of the existence of algorithms, and, under algorithmic complexity constraints, of the existence of natural evidence of properties, are virtually without precedent in the modern study of computational complexity.

Further surprises, perhaps overshadowed by those above, include

(1) For some problems one can now prove polynomial-time complexity, without having any knowledge of the degree of the bounding polynomial, and without having any effective means of computing it.

(2) Robertson-Seymour posets are a powerful tool for proving decidability. For a variety of interesting problems there are no known alternate proofs of decidability in any complexity class.

(3) The algorithms that are promised to exist have running times bounded by polynomials with hilariously astronomical constants of proportionality.

(4) There are indications that the algorithms themselves may in many cases be enormous — so large that they could not be written down in any medium in this universe.

For the most part, this paper is a survey of applications of the combinatorial finite basis theorems of Robertson and Seymour to problems of concrete computational complexity. An attempt is made to place these developments in the perspective of a broad program of research in computational complexity based on partial orders of combinatorial objects and to point to the horizons that have been opened up.

Section Two introduces six concrete problems that we will use as examples.

Section Three defines several partial orders on finite graphs based on sequences of local operations and states the Robertson-Seymour theorems.

Section Four describes five applications of the well partial ordering of graphs by minors and immersions, and an open problem.

Section Five discusses the strong minor order, and presents some new results that bear on the complexity of order testing. Many intersection-defined classes of graphs are closed in the strong minor order.

Section Six explores the practical significance of complexity results based on the Robertson-Seymour theorems. Some ways in which practical algorithms might be developed are surveyed, as well as some progress in the development of direct algorithms with similar asymptotic performance.

## 2. Six problems.

In this Section we introduce six problems of combinatorial computing that we will use throughout this paper to illustrate applications of the Robertson-Seymour theorems and the various issues that these applications raise.

### knotlessness

It has been shown [CG] that no matter how the complete graph $K_7$ is embedded in 3-space there must be a cycle in the graph that is nontrivially knotted. Of course, any planar graph has a knotless embedding in 3-space. The recognition problem is defined

Instance: A graph $G$.

Question: Does $G$ have a knotless embedding in 3-space?

There is no obvious argument (apart from the one we will employ) that the problem is even *decidable*, let alone in $NP$.

### gate matrix layout

The problem has practical applications in VLSI layout, for a variety of layout styles. It was first introduced under this name in [LL], but essentially the same combinatorial problem has appeared under other names as well, such as "PLAs with multiple folding," and "Weinberger arrays." It is closely related to the notion of *pathwidth* introduced by Robertson and Seymour in the first stage of their structure theory. The problem is defined formally

Instance: A Boolean matrix $M$ and an integer $k$.

Question: Is there a permutation of the columns of $M$ so that, if in each row we change to $*$ every 0 lying between the row's leftmost and rightmost 1, then no column contains more than $k$ 1s and $*$s?

The problem has been shown to be $NP$-complete [KF]. We are concerned here with the complexity of the problem for fixed $k$.

### feedback vertex set

The problem is well-known from [GJ] and is defined

Instance: A graph $G$ and an integer $k$.

Question: Is there a set $V' \subseteq V(G)$ with $|V'| \leq k$ and $G - V'$ acyclic?

**crossing number**

The *crossing number* of a graph $G$ is the least $k$ such that $G$ can be drawn in the plane with $k$ crossings of pairs of edges. When $k$ is part of the input the problem is $NP$-complete [GJ]. We will be concerned with the problem for fixed $k$. The general problem is defined

Instance: A graph $G$ and an integer $k$.

Question: Is the crossing number of $G$ at most $k$?

**min cut linear arrangement**

The *cutwidth* of a graph $G = (V, E)$ with respect to a permutation $f : V \to \{1, \ldots, |V|\}$, denoted $cut(G, f)$, is the least $k$ such that $|\{uv \in E : f(u) \leq i < f(v)\}| \leq k$ for $i = 1, \ldots, |V| - 1$. Equivalently, in the linear layout of $G$ described by $f$, at most $k$ edges are cut when cutting between the vertices. The *cutwidth* of $G$ is defined to be $cut(G) = min_f\ cut(G, f)$. The general problem is known to be NP-complete [GJ].

Instance: A graph $G$ and an integer $k$.

Question: Is the cutwidth of $G$ at most $k$?

**string graph**

A simple graph $G = (V, E)$ is a *string graph* if $G$ is representable by intersections of curves in the plane [EET,KGK,Si]. In other words, $V$ can be put in one-to-one correspondence with a collection of images of continuous functions from [0,1] to the plane, so that $uv \in E$ if and only if the images corresponding to $u$ and $v$ intersect. The recognition problem is defined

Instance: A graph $G$.

Question: Is $G$ a string graph?

## 3. Local operation sequences and Robertson-Seymour posets.

There are many simple local operations on combinatorial objects under which various properties of these objects might be preserved. Although our concern in this section will be with finite graphs, we note in passing that similar considerations can naturally be made for other kinds of objects, such as matroids, strings of symbols and hypergraphs.

The following are some local operations on graphs. Graphs are considered topologically and may have loops and multiple edges, and are always finite.

($I$) Delete an isolated vertex.

($E$) Delete an edge.

($V$) Delete a vertex.

($C$) Contract an edge, topologically. (This may create loops or multiple edges.)

($C'$) Contract an edge, combinatorially. (Remove $uv$, identify $u$ and $v$ and make the resulting vertex adjacent to all vertices $x \in N(u) \cup N(v)$.)

($T$) Contract an edge incident to a loopless vertex of degree two. Equivalently, remove a subdivision.

($L$) Lift an edge. Replace a pair of edges incident on $\{u, v\}$ and $\{v, w\}$, respectively, with an edge incident on $\{u, w\}$. (No assumption is made that these are distinct vertices.)

($F$) Fracture a vertex. Replace $v$ with two new vertices $v_1$ and $v_2$ and partition the edges incident on $v$ into two classes, making these incident, respectively, on $v_1$ and $v_2$.

($S$) Simplify. Delete an edge of which there is more than one copy, or remove a loop.

A family of graphs $\mathbf{F}$ is *closed* under a local operation $O$ if $G \in \mathbf{F}$ implies that for every graph $H$ obtained from $G$ by $O$, $H \in \mathbf{F}$. In the examples that follow, closure under the operations listed implies in some cases closure under other of the operations defined above as well.

**Examples**

1. The family of knotless graphs is closed under the operations $I$, $E$ and $C$.
2. The family of graphs that have a $k$-element feedback vertex set is closed under the operations $I$, $E$ and $C$.
3. The family of graphs with crossing number at most $k$ is closed under the operations $I$, $E$ and $T$.
4. The family of graphs of cutwidth at most $k$ is closed under the operations $I$, $E$, $T$ and $F$. (Note that closure under $T$ and $F$ implies closure under $L$.)
5. The family of string graphs is closed under the operations $V$ and $C'$.

A *quasiorder* is a relation that is reflexive and transitive. Given a collection $\mathbf{C}$ of local operations we can define a quasiorder in a natural way by setting $G \geq_{\mathbf{C}} H$ if and only if a graph isomorphic to $H$ can be obtained from $G$ by a sequence of local operations chosen from **C**.

In general, a collection of local operations yields a quasiorder in this way, but in the examples below the quasiorder is also antisymmetric, i.e., a partial order.

**Examples**

1. If $\mathbf{C} = \{I, E, C\}$ then $\geq_{\mathbf{C}}$ is termed the *minor* order and is denoted $\geq_m$.
2. If $\mathbf{C} = \{V, C, S\}$ then $\geq_{\mathbf{C}}$ is termed the *strong minor* order and is denoted $\geq_{sm}$ .
3. If $\mathbf{C} = \{V, C'\}$ then $\geq_{\mathbf{C}}$ is termed the *combinatorial strong minor* order and is denoted $\geq_{csm}$
4. If $\mathbf{C} = \{I, E, L\}$ then $\geq_{\mathbf{C}}$ is termed the *immersion* order and is denoted $\geq_i$.
5. If $\mathbf{C} = \{I, E, F, T\}$ then $\geq_{\mathbf{C}}$ is termed the *weak immersion* order and is denoted $\geq_{wi}$.
6. If $\mathbf{C} = \{I, E, T\}$ then $\geq_{\mathbf{C}}$ is termed the *topological* order and is denoted $\geq_t$.

The above list is just a small sample of the interesting collections of local operations. It does include some that are widely applicable and for which we now have general information regarding the computational complexity of membership in closed families of graphs. With respect to many such orders, much remains unknown.

In a lengthy series of papers developing a deep and beautiful structure theory, Robertson and Seymour have proven (or announced) the following.

**Theorem 1** [RS6] Any set of finite graphs contains only a finite number of minimal elements in the minor order.

**Theorem 2** [RS5] For every fixed graph $H$, the problem that takes as input a graph $G$ and determines whether $G \geq_m H$ is solvable in polynomial time.

**Theorem 3** [RS4] Any set of finite graphs contains only a finite number of minimal elements in the immersion order.

**Theorem 4** [FL3] For every fixed graph $H$, the problem that takes as input a graph $G$ and determines whether $G \geq_i H$ is solvable in polynomial time.

Theorem 4 is an easy consequence of results in [RS5] on the $k$ disjoint connecting paths problem.

The proof of Theorems 1 and 3 is *nonconstructive.*

Even if a set of graphs has a finite description, for example, a description of a Turing machine that accepts it, the arguments that prove Theorems 1 and 3 provide no means for computing the finite set of minimal elements that is promised to exist. It has been shown that no constructive proof of such general theorems is possible [FRS].

A family of graphs $\mathbf{F}$ is *closed* under a quasiorder $\leq$ if $G \in \mathbf{F}$ and $H \leq G$ imply that $H \in \mathbf{F}$. If $\mathbf{F}$ is closed under $\leq$ then an *obstruction set* for $\mathbf{F}$ is a subset $\mathbf{O}$ of the complement of $\mathbf{F}$, $\mathbf{O} \subseteq \overline{\mathbf{F}}$, satisyfing

1. If $H \in \mathbf{F}$ then for some $K \in \mathbf{O}$, $H \geq K$.
2. If $H \geq K$, $H \neq K$ and $K \in \mathbf{O}$ then $H \notin \mathbf{O}$.

An obstruction set $\mathbf{O}$ characterizes $\mathbf{F}$ in the following way.

**Lemma 1** If $\mathbf{O}$ is an obstruction set for $\mathbf{F}$ in the quasiorder $\geq$, then $H \in \mathbf{F}$ if and only if it is not the case that $H \geq K$ for some $K \in \mathbf{O}$. □

If the quasiorder $\geq$ is actually a partial order then a closed family $\mathbf{F}$ has a unique obstruction set consisting of the minimal elements of $\overline{\mathbf{F}}$.

Theorems 1-4 and an additional argument [RS5] establish that any family of graphs closed in the minor ordering has an $O(n^3)$ membership test. If also there is a planar graph not in the family then there is an $O(n^2)$ membership test. A family of graphs closed in the immersion order has an $O(n^2)$ membership test if there is a planar graph of maximum degree 3 not in the family. The Robertson-Seymour theory is presently less well-developed for the immersion order, as it was discovered only later (see [RS1]) that their approach would work for immersions as well as for minors.

## 4. Five Consequences.

This Section surveys how the Robertson-Seymour theorems can be applied to obtain, from the complexity standpoint, very interesting results about our sample of problems.

**Consequence 1** [FL2] Knotless graphs can be recognized in time $O(n^3)$. □

The proof is trivial. We have observed that the family $\mathbf{F}$ of knotless graphs is closed under the operations $I, E$ and $C$, and therefore $\mathbf{F}$ is closed in the minor ordering. By

Theorem 1, $\overline{\mathbf{F}}$ has a finite number of minimal elements $\mathbf{O}$ that characterize $\mathbf{F}$. For an arbitrary graph $G$ we can determine if $G$ is knotless by testing, for each graph $H \in \mathbf{O}$, whether $G \geq_m H$, using the order tests of Theorem 2. A similar application to linkless embedding of graphs in 3-space was noted by Nešetřil and Thomas [NT].

Because the proof of Theorem 1 is nonconstructive, we do not know what the obstruction set $\mathbf{O}$ for knotlessness is, and we do not know how to find it — we know only that it is *finite* — and so we do not know the algorithm. Moreover, even if the obstruction set and therefore the algorithm were known, it would not tell us what a knotless embedding of $G$ might be, but only whether one is possible. At present, no alternate proof that knotlessness is decidable has been found, despite some consideration by experts in low-dimensional topology.

**Consequence 2** [FL1] For every fixed positive integer $k$, the $k$-gate matrix layout problem can be solved in time $O(n^2)$. □

This is established by a combinatorial reduction of the matrix problem to a problem about graphs. Each matrix $M$ is mapped to a graph $f(M)$, and the mapping $f$ has the following properties, where $\mathbf{M_k}$ is the set of all "yes" instances to the problem.

1. $f(\mathbf{M_k})$ is $\geq_m$ closed.
2. For every graph $G$, either $f^{-1}(G) \subseteq \mathbf{M_k}$ or $f^{-1}(G) \cap \mathbf{M_k} = \emptyset$.

There is the possibility that reductions of this sort might be useful for many kinds of problems (not necessarily about graphs or hypergraphs). Surprisingly, the families of graphs $f(\mathbf{M_k})$ are precisely the graphs of pathwidth [RS3] at most $k-1$. The proof of this will appear elsewhere. It seems likely that the dynamic programming approach of Sudborough, et. al., to problems of various "width" metrics [EST,MPS,MaS] could be used to show, constructively, that the problem is solvable in time $O(n^{O(k)})$.

**Consequence 3** For every fixed positive integer $k$, the problem $k-$feedback vertex set can be solved in time $O(n^2)$. □

One simply observes that the relevant families of graphs are closed in the minor order, and that there are obvious planar obstructions (such as the disjoint union of $k+1$ copies of $K_3$). The problem is clearly solvable constructively in time $O(n^k)$ by trying all $k-$subsets of the vertex set.

**Consequence 4** [FL1] For every fixed positive integer $k$, it can be determined in time $O(n^3)$ whether a graph $G$ of maximum degree three has crossing number at most $k$. □

As we noted in Section 3, the family of graphs $\mathbf{F}$ with crossing number at most $k$ is closed under the operations $I$, $E$ and $T$. But it is not closed under the operations $C$ or $L$, and therefore it is not closed in either the minor or immersion orders. It is closed under the topological order $\leq_t$, but this is not a well partial order, and the obstructions sets are infinite.

However, if we restrict our attention to graphs of maximum degree three, then the orders on these graphs inherited from the ordering of all graphs under $\leq_m$ and $\leq_t$ coincide. Thus, by Theorem 1, there are at most a finite number of obstructions in the topological ordering, for graphs of maximum degree three. A theorem of Mader [Ma] states that for every positive integer $r$, the set of graphs not containing $r$ disjoint cycles is well partially ordered by $\geq_t$, and this allows us to claim a similar Corollary for this restriction on input, in this case with running time $O(n^2)$.

**Consequence 5** [FL3] For every fixed positive integer $k$ the problem of $k-$min cut linear arrangement is solvable in time $O(n^2)$. □

The relevant graph families $\mathbf{F_k}$ are closed in the immersion order, but not in the minor order. This only guarantees decidability in time bounded by some undetermined polynomial of degree that depends on the order of the largest obstruction.

There are two ways in which this can be improved. First, the general problem reduces combinatorially to the problem for graphs of maximum degree three [MoS2], with quadratic blow-up in the order of the graph. Since $\mathbf{F_k}$ is $\geq_t$ closed, this yields an $O(n^4)$ decision algorithm.

A second approach is based on the following easy Lemma.

**Lemma 2** Let $H$ be a graph of maximum degree 3. If $G \geq_m H$ then $G \geq_i H$. □

A sufficiently large complete binary tree $T$ has cutwidth greater than $k$, so by Lemma 2, the graphs of the family $\mathbf{F_k}$ of "yes" instances of the problem exclude $T$ as a minor, and so, by a theorem of Robertson and Seymour [RS3], have bounded treewidth (in fact, bounded pathwidth). For graphs of bounded treewidth, testing for the $\mathbf{F_k}$-obstructions in the immersion order can be done in time $O(n^2)$.

String graphs are closed under $\leq_{csm}$ but not under $\leq_m, \leq_i$ or $\leq_t$ and the complexity of recognizing string graphs remains open, even for graphs of maximum degree three.

The most exciting aspect of Theorems 1-4 is that they constitute a new paradigm for determining the complexity of problems.

## 5. Strong minors.

Many intersection defined families of simple graphs, including interval graphs, circular arc graphs, chordal graphs, trapezoid graphs [CK,DGP] and string graphs — are closed in the combinatorial strong minor order. This suggests the possibility of a systematic approach to the complexity of recognizing such families.

Thomas [Th] has shown that $\geq_{csm}$ is not a well partial ordering of simple graphs (and thus $\geq_{sm}$ is not a well partial ordering of general graphs) by exhibiting an infinite antichain of planar graphs.

Even so, many of the graph families mentioned above have finite obstruction sets in the combinatorial strong minor order. For example,

**Theorem 5** A simple graph $G$ is not an interval graph if and only if

$G \geq_{csm} H$ for some $H$ pictured below

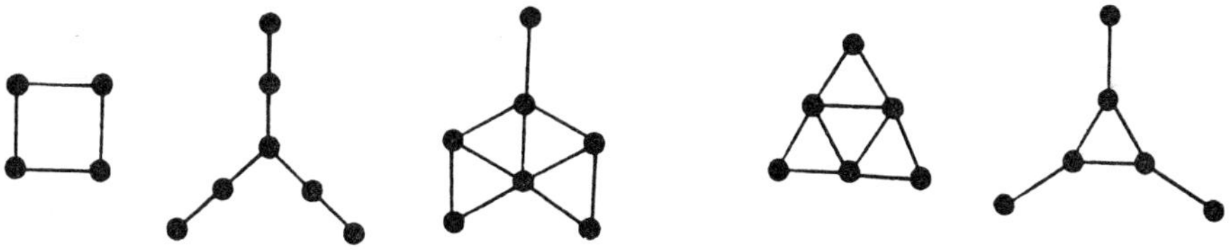

**Proof** The Lekerkerker-Boland theorem [LB] identifies an infinite obstruction set (of a few types of graphs) in the stronger partial ordering of simple graphs by induced subgraphs. It is easy to see that we need only find the minimal elements of the Lekerkerker-Boland obstruction set in the order $\geq_{csm}$, and these are the graphs pictured. □

Similarly, a graph $G$ is not chordal if and only if $G \geq_{csm} C_4$.

Although $\geq_{sm}$ and $\geq_{csm}$ are not well partial orders in general, we might hope for a positive result in a restricted setting. Thomas has shown the following result, which is in some sense close to the best possible.

**Theorem 6** [Th] The order $\geq_{csm}$ is a well partial order of simple series-parallel graphs. □

This result can be "lifted" to the setting of general (topologically considered) graphs by a combinatorial reduction. The argument will appear elsewhere.

**Theorem 7** The order $\geq_{sm}$ is a well partial order of series-parallel graphs. □

It is presently not known whether there is any graph $H$ for which the problem of order testing in the combinatorial strong minor order is *NP*- complete. Trivially, since $G \geq_{csm} K_n$ if and only if $G \geq_m K_n$ for simple graphs $G$, there are polynomial-time order tests for $H$ a complete graph.

One possible approach to the problem of order tests for the combinatorial strong minor order would be to imitate the original arguments of Robertson and Seymour that the minor order has polynomial-time order tests. They were able to exploit a reduction of the problem of minor testing to the $k$-disjoint connecting paths problem [RS1]. Similarly, the problem of order testing in the combinatorial strong minor order can be reduced to the following analogous problem.

**$k$-induced connecting paths**

Instance: A graph $G$, and $k$ pairs of vertices $(s_i, t_i)$, $i = 1, \ldots, k$.

Question: Is there a vertex induced subgraph of $G$ that consists of $k$ paths joining $s_i$ to $t_i$ for $i = 1, \ldots, k$?

In contrast to the $k$-disjoint connecting paths problem that can be solved in time $O(n^3)$ for all $k$ [RS5], we have the following negative result.

**Theorem 8** The $k-$induced connecting paths problem is *NP*-complete for $k = 2$.

**Proof** The problem is clearly in *NP*. The reduction is from $3SAT$. A simple example of the construction is pictured in Figure 2.

A variable component consists of one of the units shown in Figure 1.

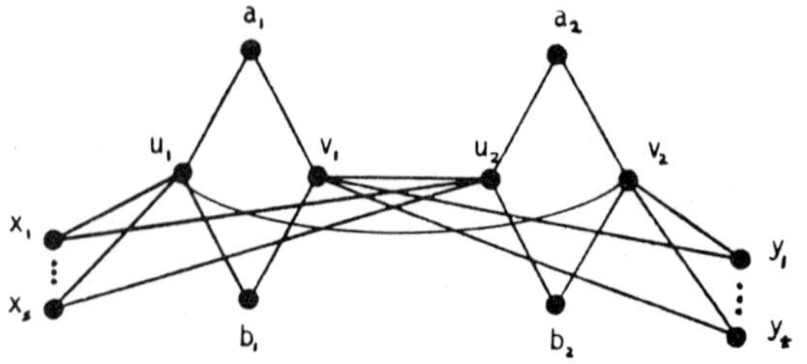


Figure 1
A variable component.

Suppose that there are induced paths $\rho_1$ and $\rho_2$ in a variable component that each have length at least 2 and that have, respectively, $a_1$ and $a_2$ as endpoints. Then it is easy to check that the only possibilities are

(i) $\rho_1 = (a_1, u_1, b_1)$, $\rho_2 = (a_2, u_2, b_2)$

or

(ii) $\rho_1 = (a_1, v_1, b_1)$, $\rho_2 = (a_2, v_2, b_2)$

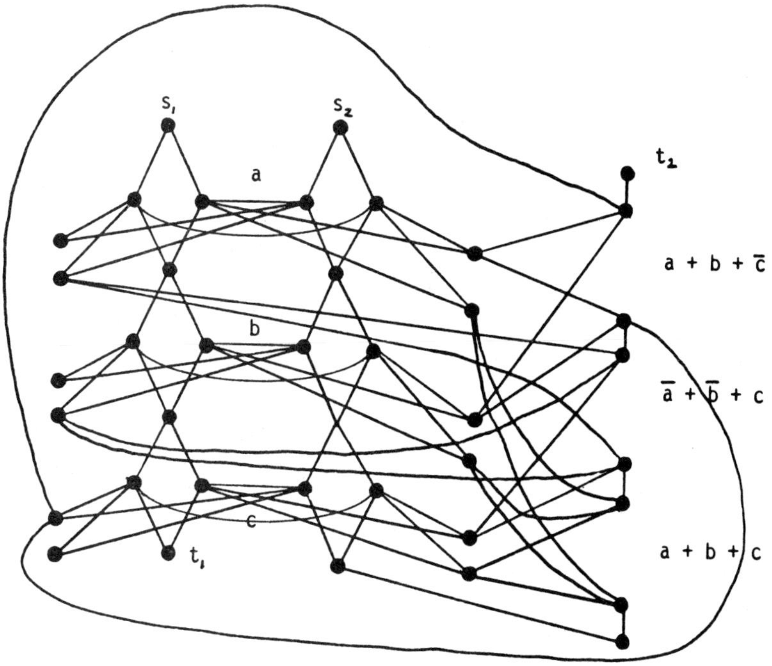


Figure 2
An example of the construction $E = (a + b + \bar{c})(\bar{a} + \bar{b} + c)(a + b + c)$.

Suppose $E$ is satisfiable by a truth assignment $\tau$, and suppose $\tau(a) = true$ for a variable $a$. Then in the graph $G_E$ the path $\rho_1$ should include, in the variable component that corresponds to $a$, the subpaths (i). The subpaths (ii) should be included if $\tau(a) = false$. It is clear how the second induced path $\rho_2$ can be routed. The "gap" corresponding to a clause $C$ can be crossed via a vertex $x_i$ or $y_j$ corresponding to a true literal of $C$.

Conversely, if there is an induced subgraph consisting of the two required paths, then by considering how the components are attached in $G_E$, one can read off a truth assignment from $\rho_1$. The path $\rho_2$ cannot contain any vertices of any variable component other than vertices $x_i$ or $y_j$. If a gap corresponding to a clause $C$ can be crossed then one of the literals of $C$ evaluates to true in the truth assignment corresponding to $\rho_1$. □

## 6. Toward practical algorithms

What is the practical significance of the Robertson-Seymour theorems? At the very least they provide tools that can be used to provide complexity bounds (and decidability results) that the algorithm designer can take as guidance and encouragement. But also there are other more direct ways that practical algorithms might be developed from the Robertson-Seymour structure theory. In this Section we review some of these possibilities with respect to our sample of problems.

The most hopeful cases are those problems for which there are planar "no" instances. This does not include the problem of recognizing knotless graphs, since every planar graph clearly has a knotless embedding in 3-space, and it does not include the problem of recognizing graphs with crossing number at most $k$. There is almost nothing to say about these problems, at present, with respect to the development of practical algorithms. Crossing number can be solved constructively by a trivial exhaustive algorithm in time $O(n^{2k})$, but no constructive algorithm to decide knotlessness is known, nor is there known any self-reduction algorithm to produce a knotless embedding, when one exists, using the decision problem as an oracle. The only known obstruction is $K_7$ [CG].

The problems of $k$-gate matrix layout, $k$-min cut linear arrangement and $k$-feedback vertex set do have planar "no" instances. There are four difficulties to be overcome.

(1) In applying Theorems 1-4 enormous constants are introduced in bounding the treewidth of the "yes" instances. (See [Jo] for a discussion of this.)

(2) There might also be enormous constants introduced by the size of the obstruction sets.

(3) The obstruction sets are not known (the theorems only assert that they are finite).

(4) Knowing the decision algorithm might not provide a solution to the construction, or search problem.

The constant $\theta(k)$ that would be introduced in (1) for all three of these problems would make the universe quite uncomfortable even for $k = 2$, but this is really only an artifact of the great generality of the Theorems. In all three cases a very reasonable bound can be proved by considerations particular to the problem. In the case of $k$- gate matrix layout the bound is $k - 1$. For several such results see [Bo].

For many problems, including $k$-feedback vertex set, the three remaining problems are moot in the bounded tree-width setting. A variety of related approaches [AP,BLW,

MoS1,Se,WHL,Wi] have shown that a great many problems, including many that are $NP$-complete for unrestricted input, are solvable constructively (natural evidence is produced) in linear-time for graphs of bounded tree-width for which a tree-decomposition is supplied as part of the input. An "approximate" tree decomposition can be found in quadratic time by an algorithm of Robertson and Seymour [RS5]. More careful work remains to be done to adapt and refine the techniques cited to make them as efficient as possible. The main point is that in this bounded tree- (or branch-) width setting one may in many cases simply solve the problem directly, instead of looking for the obstructions. For the case of $k$-feedback vertex set this gives a constructive $O(n^2)$ algorithm, with hidden constant approximately $(3k)^{3k}$. A different approach, inspired by the Robertson-Seymour complexity bound and based on fast self-reduction, has yielded a constructive $O(n^2)$ algorithm with hidden constant approximately $(2k)^k$ [F,FL4]. Perhaps randomized algorithms can provide additional speed-up for some of these problems.

Unfortunately for the sake of our other two sample problems, it is presently not known how to solve vertex permutation problems directly in the bounded tree-width setting (the techniques cited above seem not to apply to these problems [Wi]), and so we must look for obstructions. In the case of $k$-gate matrix layout, there are two obstructions to layout cost 2 [FL1]. A set of 110 obstructions to layout cost 3 has been identified and it is believed to be complete [BFKL]. For layout cost 4 there are at least 1.3 million obstructions, and in general the number of obstructions seems to grow as an exponential tower of height linear in $k$, providing an illustration of problem (2).

A consideration of the obstruction set $\{K_{3,3}, K_5\}$ for planarity suggests an approach. The single graph $K_{3,3}$ is a very good approximation to the whole set. It may be that for some problems a small subset of the obstruction set contains most of the information, and that useful algorithms can be developed on the basis of *approximate* obstruction sets. A little experience with gate matrix layout seems to support this possibility [FKL].

Another possibility is that by considering a weaker order or quasiorder under which the relevant family of graphs is closed, the number of obstructions might be reduced, while still supporting linear-time order tests in the bounded branchwidth setting. There are 15 obstructions to cutwidth 2 in the immersion order, while there are 5 in the weak immersion order [MaS].

It is likely to be many years before the practical significance of the Robertson-Seymour theorems is fully understood, and the many research horizons opened up by these results have been explored.

# References


[AP] S. Arnborg and A. Proskurowski, "Linear Time Algorithms for NP-hard Problems on Graphs Embedded in $k$-trees," TRITA-NA-8404, The Royal Institute of Technology (1984).

[Be] M. J. Beeson, Foundations of Constructive Mathematics, Springer-Verlag, New York, 1980.

[BFKL] R. L. Bryant, M. R. Fellows, N. G. Kinnersley and M. L. Langston, "On Finding Obstruction Sets and Polynomial-Time Algorithms for Gate Matrix Layout," *Proc. 25th Allerton Conf. on Communication, Control, and Computing* (1987).

[BLW] M. Bern, E. Lawler and A. Wong, "Why Certain Subgraph Computations Require Only Linear Time," *Proc. 26th Symp. on the Found. of Computer Science*, 1985.

[Bo] H. L. Bodlaender, "Classes of Graphs with Bounded Tree-Width," Technical Report RUU-CS- 86-22, Department of Computer Science, University of Utrecht, 1986.

[BR] D. Bridges and F. Richman, Varieties of Constructive Mathematics, Cambrdige Univ. Press, New York, 1987.

[CG] J. H. Conway and C. Gordon, "Knots and Links in Spatial Graphs," *J. Graph Th.* 7 (1983), 445–453.

[CK] D. Corneil and P. Kamula, "Extensions of Permutation and Interval Graphs," to appear.

[DGP] I. Dagan, M. Golumbic and R. Pinter, "Trapezoid Graphs and Their Coloring," TR 88-206, November 1986, IBM Israel.

[EET] G. Ehrlich, S. Even and R. Tarjan, "Intersection Graphs of Curves in the Plane," *J. Comb. Th. Ser. B* 21 (1976), 8–20.

[EST] J. Ellis, I. Sudborough and J. Turner, "Graph Separation and Search Number," to appear.

[F] M. R. Fellows, "Achieving the Robertson-Seymour Bounds Constructively: $k$-Feedback and Related Vertex and Edge Sets," to appear.

[FL1] M. R. Fellows and M. A. Langston, "Nonconstructive Advances in Polynomial-Time Complexity," *Info. Proc. Letters*, 26 (1987/88) 157–162.

[FL2] ______, "Nonconstructive Tools for Proving Polynomial- Time Decidability," *J. of the ACM*, to appear.

[FL3] ______, "Layout Permutation Problems and Well-Partially-Ordered Sets," *Proc. 5th MIT Conf. on Advanced Research in VLSI*, 1988, to appear.

[FL4] ______, "Fast Self-Reduction Algorithms for Combinatorial Problems of VLSI Design," to appear.

[FKL] M.R. Fellows, N.G. Kinnersley and M.A. Langston, "An Automated Learning System for Layout Optimization," to appear.

[FRS] H. Friedman, N. Robertson and P. D. Seymour, "The Metamathematics of the Graph Minor Theorem," in Applications of Logic to Combinatorics, AMS Contemporary Mathematics Series, American Math. Soc., Providence, RI, to appear.

[GJ] M. R. Garey and D. S. Johnson, Computers and Intractability: A Guide to the Theory of NP- Completeness, Freeman, San Francisco, CA, 1979.

[Jo] D. S. Johnson, "The Many Faces of Polynomial Time," in the NP-Completeness Column: An Ongoing Guide, *J. Algorithms* 8 (1987), 285–303.

[KF] T. Kashiwabara and T. Fujisawa, "NP- completeness of the Problem of Finding a Minimum-Clique-Number Interval Graph Containing a Given Graph as a Subgraph," *Proc. IEEE Symp. on Circuits and Systems* (1979), 657–660.

[KGK] J. Kratochvíl, Goljan and P. Kučera, "String Graphs," Rozpravy Československé Akademie Věd, Prague, 1986.

[LB] C. Lekkerkerker and J. Boland, "Representation of a Finite Graph by a Set of Intervals on the Real Line," *Fundam. Math.* 51 (1962), 45–64.

[LL] A. Lopez and H. Law, "A Dense Gate Matrix Layout Method for MOS VLSI," *IEEE Trans. Elec. Devices 27* (1980), 1671–1675.

[Ma] W. Mader, "Wohlquasigeordnete Klassen endlicher Graphen," *J. Combinatorial Theory (Ser. B)*, 12 (1972), 105–122.

[MaS] F. S. Makedon and I. H. Sudborough, "On Minimizing Width in Linear Layouts," to appear.

[MoS1] B. Monien and I.H. Sudborough, "Bandwidth Constrained NP-Complete Problems," *Theoretical Computer Science* 41 (1985), 141– 167.

[MoS2] ______, "Min Cut is NP-Complete for Edge Weighted Trees," to appear.

[MPS] F. S. Makedon, C. H. Papadimitrious and I. H. Sudborough, "Topological Bandwidth," *SIAM J. Alg. Disc. Meth* 6 (1985), 418–444.

[NT] J. Nešetřil and R. Thomas, "A Note On Spatial Representations of Graphs," *Comm. Math. Univ. Carol.* 26 (1985) 655–659.

[Re] C. Reid, Hilbert, Springer-Verlag, New York, 1970.

[RS1] N. Robertson and P. Seymour "Disjoint Paths – a Survey," *SIAM J. Alg. Disc. Meth.* 6 (1985), 300–305.

[RS2] ______, "Graph Minors – a Survey," in Surveys in Combinatorics (I. Anderson, ed.) Cambridge Univ. Press (1985), 153–171.

[RS3] ______, "Graph Minors I. Exluding a Forest," *J. Combinatorial Theory (Ser. B)* 35 (1983), 39–61.

[RS4] ______, "Graph Minors IV. Tree-Width and Well-Quasi- Ordering," to appear.

[RS5] ______, "Graph Minors XIII. The Disjoint Paths Problem," to appear.

[RS6] ______, "Graph Minors XVI. Wagner's Conjecture," to appear.

[Se] D. Seese, "Tree-Partite Graphs and the Complexity of Algorithms," Preprint P-Math- 08/86, Karl-Weierstrass-Institut fur Mathematik, Akademie der Wissenschaften der DDR, 1986.

[Si] F. Sinden, "Topology of Thin Film RC Circuits," *Bell Sys. Tech. J.* 45 (1966) 1639–1662.

[Th] R. Thomas, "Graphs Without $K_4$ and Well-Quasi-Ordering," *J. Comb. Th. Ser. B* 38 (1985) 240–247.

[Wi] T. V. Wimer, "Linear Algorithms on $k$-Terminal Graphs," Ph.D. Dissertation, Clemson University, 1987.

[WHL] T. V. Wimer, S.T. Hedetniemi and R. Laskar, "A Methodology for Constructing Linear Graph Algorithms," *Cong. Num.* 50 (1985) 43–60.

COMPUTER SCIENCE DEPARTMENT
UNIVERSITY OF IDAHO
MOSCOW, IDAHO 83843

Contemporary Mathematics
Volume 89, 1989

# ON GENUS-REDUCING AND PLANARIZING ALGORITHMS FOR EMBEDDED GRAPHS

Joan P. Hutchinson[1]

ABSTRACT. This paper surveys what is known concerning the problems of finding small sets of vertices in an embedded graph whose removal reduces the genus or whose removal leaves a planar graph. These results are related to the separator theorem for embedded graphs, which guarantees a small set of vertices whose removal leaves all components small.

1. INTRODUCTION. This paper considers two questions in topological graph theory.

QUESTION 1. *Given a graph, how small a set of vertices is there whose removal leaves a planar graph?*

The easiest way to solve this problem, if we ignore the "smallness" requirement, is to remove all but four vertices of the graph: the remaining graph is surely planar; however, most likely little of the structure of the initial graph has been retained. Another efficient approach is to try one of the planarity-testing algorithms [**9**,**12**], which run in time linear in the number of edges of the graph. These proceed by adding to an initial planar subgraph until an edge crossing occurs; however, there is no guarantee that the last planar subgraph is a maximum. It is an NP-complete problem to determine the size (number of vertices) of a maximum induced planar subgraph of a nonplanar graph [**6**, p. 195]; Question 1 asks for the size of the complementary set of vertices.

QUESTION 2. *Given a graph, how small a set of vertices is there whose removal reduces the genus of the graph?*

Clearly a solution of this question leads to a solution of the first by repeated application; however, as we shall see, the best known results for the first question do not come from results for the second, but in fact the opposite is true. The following is known about Question 2.

---

1980 *Mathematics Subject Classification* (1985 *Revision*). 05C10, 68Q25
[1] Supported by NSF Grant No. DCR-8411690 and the Mathematical Sciences Research Institute, Berkeley, Calif.

THEOREM 1. [**1**] *If G is a nonplanar graph with n vertices, then there is a set of at most* $\sqrt{2n}$ *vertices whose deletion reduces the genus of G.*

In fact, if the graph is embedded as a triangulation of a surface, then there is a noncontractible and nonseparating cycle (see definitions below) that is such a small genus-reducing set. The proof of this fact is existential, rather than algorithmic, and requires an embedding of the graph.

Here are the basic definitions we use; see also [**2**,**8**,**15**]. A graph is said to *embed* on a surface of genus $g \geq 0$ if it can be drawn on the sphere with $g$ handles, denoted $S(g)$, so that no two edges cross; in this paper we consider embeddings only on these orientable surfaces. The *genus* of a graph $G$ is the least integer $g$ for which $G$ embeds on $S(g)$. A *face* of an embedding of $G$ on $S(g)$ is a connected component of $S(g)\setminus G$ and is called a *2-cell* if it is homeomorphic to the (open) interior of a disc. An embedding is called a *2-cell embedding* if every face is a 2-cell and a *triangulation* if every face is bounded by three edges. A cycle of $G$ is called *contractible* if it is contained within the interior of a disc; otherwise it is called *noncontractible*. The crucial result for embedded graphs follows.

EULER'S FORMULA. *If G has a 2-cell embedding on S(g),* $g \geq 0$, *then* $n - e + f = 2 - 2g$, *where n, e, and f are, respectively, the number of vertices, edges, and faces of the embedded graph.*

In this work we shall always assume that given a graph, we are also given an embedding of the graph on a surface of known genus. It is not hard to find an embedding of a graph on some surface, but the problem of determining the surface of minimum genus on which a given graph embeds is one of unknown complexity. Like the graph isomorphism problem [**6**, p. 285], the genus problem is one of the few remaining open problems, listed in [**6**], for which there is neither a polynomial-time algorithm nor a proof that the problem is NP-complete. The best result in this direction is the following.

THEOREM 2. [**5**] *Given an n-vertex graph G and a positive integer g there is an algorithm that runs in time* $O(n^{O(g)})$ *to determine whether G has genus at most g.*

Thus for constant or bounded genus, this is a polynomial algorithm.

Some interesting progress has been made recently by C. Thomassen [**14**], who has found a polynomial-time algorithm for embedding a large class of graphs: a graph is said to have a *large-edge-width (LEW) embedding* if it can be embedded so that the length (i.e., number of edges) of the largest face boundary is smaller than the length of the shortest noncontractible cycle. Then Thomassen has found an algorithm to determine whether or not a given 3-connected graph has an LEW embedding, and if so, to find the embedding, which he proves is the unique embedding of the graph on its genus surface.

In this paper we consider the easier situation in which we are given a graph and an embedding on a surface of genus $g > 0$. Then quite a bit more can be said about the questions posed above. Regarding Question 1 it is easily seen that in an $n$-vertex graph of genus $g$, there is a set of $O(g\sqrt{n})$ vertices whose removal leaves a planar graph by applying the result of Theorem 1 $g$ times. In [**7**] it was conjectured that the correct bound was $O(\sqrt{gn})$; this bound is proved in [**11**].

THEOREM 3. *If $G$ is a graph with $n$ vertices and genus $g$, then it contains a set of at most $26\sqrt{gn}$ vertices whose removal leaves a planar graph.*

This result is best possible, up to constants, since there are graphs with $n$ vertices and of genus at most $g$ for which there is no $o(\sqrt{gn})$ "planarizing" set. The proof is a generalization of the proof of the following "genus separator" theorem.

THEOREM 4. [**3, 7**] *In an $n$-vertex graph of genus $g$, $g>0$, there is a set of $O(\sqrt{gn})$ vertices whose removal leaves all components with at most $2n/3$ vertices.*

The latter is an extension of the very useful "planar separator" theorem.

THEOREM 5. [**13**] *In an $n$-vertex planar graph there is a set of $O(\sqrt{n})$ vertices whose removal leaves all component with at most $2n/3$ vertices.*

Up to constants both separator theorems are best possible, and given embedded graphs there are algorithms to find the separating sets in time linear in the number of edges of the graph.

Regarding Question 2 the following is known.

THEOREM 6. [**10**] *Every triangulation of a surface of genus $g>0$ with $n$ vertices contains a noncontractible cycle of length at most $34\sqrt{n/g}\log g$.*

We conjecture that every triangulation contains a noncontractible cycle of length $O(\sqrt{n/g})$. The intuition for the conjecture is that if the vertices of an embedded graph $G$ were distributed evenly with $n/g$ vertices "per handle," then since there is a $O(\sqrt{n})$ noncontractible cycle in a triangulation of the torus by Theorem 1, there should be a $O(\sqrt{n/g})$ noncontractible cycle in $G$.

2. ALGORITHMS. We summarize what is known about the algorithmic implementation of Theorems 3, 4, and 6. Perhaps the most important idea in these algorithms about embedded graphs is that every step can be implemented completely combinatorially, without appeal to topology.

First, what does it mean to be "given an embedded graph" in an algorithm? If $G$ is embedded on an orientable surface $S(g)$, then a coherent orientation can be given to the surface, which means, for example, that we can assign a clockwise orientation to every vertex $v$ of $G$, giving a cyclic permutation of the edges incident with $v$. Conversely, given a graph $G$, first we may replace every edge of $G$ by two directed edges, one in each direction. Then let $\pi$ be a permutation of the directed edges of $G$ such that for each vertex $v$, $\pi$ restricted to edges of the form $(v, w)$ is a cyclic permutation of length equal to the degree of $v$. Then $\pi$ determines a unique 2-cell embedding of $G$ on some $S(g)$, which we call the *π-embedding* of $G$. This result is known as Edmonds' permutation technique [**4**].

All other information about the embedding of $G$ is recoverable from the permutation $\pi$: Define $\pi'$ on the directed edges of $G$ by $\pi'(v, w) = (x, v)$ where $\pi(v, w) = (v, x)$. Then the orbits of $\pi'$ give the faces of $G$. If $\pi'$ has $f$ orbits — and therefore the embedding has $f$ faces — then the genus $G$ of the embedding surface is available from Euler's Formula, $g = \frac{1}{2}(2 - v + e - f)$. The following additional information is obtainable and is needed in the algorithms given below:

(a) for each vertex, a cyclic list of the faces with which it is incident;

(b) for each edge, the one or two faces with which it is incident; and

(c) for each face, a cyclic list of incident vertices and edges.

In [**7**] appropriate data structures for storing this data are described.

In [**7**] the combinatorial details of the "separator" algorithm of Theorem 4 are given. The proofs of Theorems 3 and 6, like that of Theorem 4, are constructive, and lead to polynomial-time algorithms to find a small planarizing set and a short non-contractible cycle; however, they contain many technical details and are not particularly illuminating. We choose instead to describe some recent work of Thomassen to find the shortest noncontractible cycle in an embedded graph and to apply that to find a small planarizing set. His approach is also completely combinatorial, and his algorithm is implemented using the techniques described above.

THEOREM 7. [**14**] *There is a polynomial-time algorithm to find the shortest noncontractible cycle in an embedded graph.*

Before we present this algorithm, we review a few more algorithmic details about life on surfaces.

Suppose that $G$ is a $\pi$-embedded graph. As in [**14**] we call a cycle *π-separating* if its deletion from the surface leaves at least two surface components. (Note that a $\pi$-separating cycle may or may not be a separating cycle of $G$.) Then in a $\pi$-embedded graph there are three fundamental types of (simple) cycles: contractible, noncontractible and $\pi$-separating, and noncontractible and $\pi$-nonseparating. A noncontractible and $\pi$-nonseparating cycle can be pictured as wrapping around a handle; its deletion decreases the genus of the graph. A noncontractible and $\pi$-separating cycle divides the surface into surfaces of genus $g_1$ and $g_2$ where $g_1 + g_2 = g$, $g_1, g_2 > 0$; the graph is divided by the deletion of such a cycle into subgraphs of genus at most $g_1$ and $g_2$. A contractible cycle is necessarily $\pi$-separating and divides the surface into surfaces of genus 0 and $g$.

It is not hard to distinguish algorithmically between these three types of cycles. The embedding $\pi$ induces an orientation of any cycle $C$, and by traversing $C$ in that direction, one may label each edge incident with $C$, but not on $C$, as a left or right edge. (If any edge is labeled both left and right, then $C$ is $\pi$-nonseparating and hence noncontractible.) Then the vertices and edges of $C$ are replaced by two copies of $C$, one of which is incident with all left edges and the other with all right edges. In addition the two copies of $C$ are declared to be faces in the resulting embedded graph. This procedure corresponds topologically to cutting the surface along $C$, retaining copies of $C$ on each boundary, and adding in two discs. Using a connectivity algorithm, one can determine if the new graph is connected and hence $C$ was $\pi$-nonseparating and noncontractible. Otherwise $C$ was $\pi$-separating. The Euler characteristic of the resulting graphs can be determined by counting vertices, edges, and faces: $C$ was contractible or noncontractible according as one of the new graphs is or is not embedded with genus 0. This testing of $C$ can be done in time $O(e)$ where $e$ is the number of edges in the embedded graph; recall that connectivity can be determined in time $O(e)$, for example, by using a breadth-first search.

An algorithm to find the shortest noncontractible cycle in an embedded graph $G$ **[14]**.

Step 1. For each vertex x do the following:

(a) Find a breadth-first spanning tree $T_X$ of G with root x, and

(b) For each edge e not in $T_X$, find the unique cycle using e and edges of $T_X$.

Step 2. Among the cycles found in Step 1, find the shortest noncontractible cycle; this is a shortest noncontractible cycle in $G$..

This algorithm runs in time $O(ne^2)$. When applied to a triangulation, the results of Theorem 6 show that this cycle will be of length $O(\sqrt{n/g}\log g)$. However, Thomassen also proves the stronger result that the shortest noncontractible and $\pi$-nonseparating cycle found in Step 2 is the shortest such cycle in the embedded graph; by the results of Theorem 1 this is of length $O(\sqrt{n})$ in a triangulation, but presumably it is even shorter.

This algorithm can also be applied to find a small planarizing set in an embedded graph although not necessarily a smallest planarizing set. In **[7]** it was first proved that there is a set of $O(\sqrt{gn}\log g)$ vertices whose removal leaves a planar graph, and the proof leads to a recursive algorithm to find this small planarizing set. Here is an alternative algorithm.

Another algorithm to find a small planarizing set in an embedded graph $G$.

Step 1. Add edges to make $G$ a triangulation.

Step 2. Find the shortest noncontractible cycle $C$ using Thomassen's algorithm given above. Remove $C$: redefine $G$ to be $G$ minus the vertices of $C$ and all incident edges, and find the induced embedding of the remaining graph. Add edges to triangulate new faces. Then repeat Step 2 if $G$ is not embedded on the plane.

Step 3. The desired planarizing set consists of all vertices that lie on some cycle found in Step 2.

The following fact and algorithmic procedure are relevant to the implementation of Step 2. They are obvious topologically, but also are true combinatorially and algorithmically.

LEMMA. *If $G$ is embedded on a surface and $G^*$ is a subgraph of $G$, then $G^*$ receives an induced embedding from $G$.*

*Proof.* Let $\pi$ give the embedding of $G$, that is, the permutation of directed edges of $G$. At every vertex v of $G^*$, some edges of $G$ may be missing, but $\pi$ naturally induces a cyclic permutation of the remaining edges, directed out of $v$. If $\pi^*$ is the resulting permutation of directed edges of $G^*$, then $\pi^*$ gives the induced embedding. □

Algorithmically the new embedding can be obtained by constructing the new permutation $\pi^*$ from the induced permutations at each vertex, by finding the new faces from the orbits of $(\pi^*)'$, and by updating all other data. This construction and updating can be done in time $O(e)$.

THEOREM 8. *Given a $\pi$-embedded graph with $n$ vertices and $e$ edges on a surface of genus $g$, the previous algorithm finds a planarizing set of size $O(\sqrt{gn}\log g)$ in time $O(ge^2)$.*

*Proof.* The complexity bound follows because Step 2 is executed at most $O(g)$ times: At most $g-1$ times is a noncontractible and $\pi$-separating cycle found in Step 2 since each such cycle divides the surface into surfaces of genus at most $g-1$. The deletion of at most $g$ noncontractible, $\pi$-nonseparating cycles reduces the surface to the plane.

The set of vertices in Step 3 is a planarizing set even if the original graph was not a triangulation. If $C$ is not a cycle in $G$, then deleting the vertices of $C$ and incident edges in $G$ still cuts the underlying surface and either reduces the genus or separates the surface into two surfaces of smaller genus. These surfaces contain $G\backslash C$.

Finally we claim that the set of vertices collected in Step 3 has size at most $34(2\sqrt{gn}-\sqrt{n/g})\log g = O(\sqrt{gn}\log g)$ if $g>1$. The proof is by induction on $g$: It is easy to verify this result for $g=2$ since the first cycle found in Step 2 contains at most $34\sqrt{n/2}$ vertices by Theorem 6 and the remaining graphs on one or two toruses have a planarizing set of size at most $2\sqrt{2n}$ by Theorem 1.

If $g>2$ and if the first cycle $C$ found in Step 2 is $\pi$-nonseparating, then after deleting $C$ the remaining graph lies on a surface of genus $h<g$. By induction the algorithm finds a planarizing set of size at most

$$\begin{cases} 34(2\sqrt{hn}-\sqrt{n/h})\log h, & \text{if } 1<h<g \\ \sqrt{2n}, & \text{if } 1=h \end{cases}$$

$$\leq 34(2\sqrt{(g-1)n}-\sqrt{n/(g-1)})\log(g-1).$$

Then the planarizing set of Step 3 has size at most

$$34\left(2\sqrt{(g-1)n} - \sqrt{n/(g-1)}\right)\log(g-1) + 34\sqrt{n/g}\,\log g$$
$$\leq 68\sqrt{n}\,\log g\left[\sqrt{g-1} - \frac{1}{2\sqrt{g-1}} + \frac{1}{2\sqrt{g}}\right]$$
$$\leq 68\sqrt{n}\,\log g\left[\sqrt{g-1} + \frac{1}{\sqrt{g}+\sqrt{g-1}} - \frac{1}{2\sqrt{g-1}}\right]$$
$$= 68\sqrt{n}\,\log g\left[\sqrt{g} - \frac{1}{2\sqrt{g-1}}\right]$$
$$\leq 34(2\sqrt{gn} - \sqrt{n/g}\,)\log g.$$

On the other hand, if $C$, found in Step 2, is $\pi$-separating, then its removal leaves graphs embedded on surfaces of positive genus $g_1$ and $g_2$ with $g_1 + g_2 \leq g$. Suppose the graphs contain $n_1$ and $n_2$ vertices respectively where $n_1 + n_2 + d = n$ with $d = |C| \leq 34\sqrt{n/g}\log g$. By the inductive hypothesis the algorithm produces planarizing sets whose size is at most

$$34(2\sqrt{g_1 n_1} - \sqrt{n_1/g_1})\log g_1 + 34(2\sqrt{g_2 n_2} - \sqrt{n_2/g_2})\log g_2$$
$$\leq 34\left[2\sqrt{g_1 n_1} - \sqrt{n_1/g_1} + 2\sqrt{(g-g_1)(n-n_1-d)} - \sqrt{(n-n_1-d)/(g-g_1)}\right]\log g.$$

(If one or both of the surfaces is the torus, the planarizing set is even smaller.) Thus the final planarizing set has $d$ more vertices:

$$34\left[2\sqrt{g_1 n_1} - \sqrt{n_1/g_1} + 2\sqrt{(g-g_1)(n-n_1-d)} - \sqrt{(n-n_1-d)/(g-g_1)}\right]\log g + d$$
$$\leq 34\left[2\sqrt{g_1 n_1} - \sqrt{n_1/g_1} + 2\sqrt{(g-g_1)(n-n_1-d')} - \sqrt{(n-n_1-d')/(g-g_1)} + d'\right]\log g$$

where $d' = d/(34\log g) \leq \sqrt{n/g}$. In [**11**, Lemma 8] it is shown that this last expression is at most

$$34\left[2\sqrt{gn} - \sqrt{n/g}\,\right]\log g\,,$$

the desired bound of this theorem. □

In conclusion we note that the proof of Theorem 3 consists of showing that an $n$-vertex graph with a 2-cell embedding on $S(g)$, $g > 0$, contains either a $O(\sqrt{n/g})$ noncontractible cycle or else there is a sequence of planar subgraphs whose deletion leaves a graph, still with a 2-cell embedding on $S(g)$, that contains a spanning forest of radius $O(\sqrt{n/g})$ with $O(g)$ components. The resulting planarizing algorithm consists either of deleting a short noncontractible cycle and proceeding recursively, or of deleting the sequence of planar subgraphs, finding the specified spanning forest, and then proceeding as in the genus separator algorithm to find the rest of the planarizing set. This set will contain $O(\sqrt{gn})$ vertices.

The author would like to thank Carsten Thomassen and Stan Wagon for many helpful conversations on this material.

## BIBLIOGRAPHY

1. M. O. Albertson and J. P. Hutchinson, "On the independence ratio of a graph," J. Graph Theory, 2 (1978), 1–8.

2. J. A. Bondy and U. S. R. Murty, Graph Theory with Applications, American Elsevier Publishing Co., Inc., N.Y., 1976.

3. H. N. Djidjev, "A separator theorem," Comptes rendus de l'Académie bulgare des Sciences, 34 (1981), 643–645.

4. J. R. Edmonds, "A combinatorial representation for polyhedral surfaces," Notices Amer. Math. Soc., 7 (1960), 646.

5. I. S. Filotti, G. L. Miller, and J. Reif, "On determining the genus of a graph in $O(v^{O(g)})$ steps," Proc. 11th Annual ACM Symp. Theory of Computing, (1979), 27–37.

6. M. R. Garey and D. S. Johnson, Computers and Intractability, A Guide to the Theory of NP-Completeness, W. H. Freeman and Co., San Francisco, 1979.

7. J. R. Gilbert, J. P. Hutchinson, and R. E. Tarjan, "A separator theorem for graphs of bounded genus," J. Algorithms, 5 (1984), 391–407.

8. J. L. Gross and T. W. Tucker, Topological Graph Theory, John Wiley & Sons, N.Y., 1987.

9. J. E. Hopcroft and R. E. Tarjan, "Efficient planarity testing," J. Assoc. Comput. Mach., 21 (1974), 271–283.

10. J. P. Hutchinson, "On short noncontractible cycles in embedded graphs," SIAM J. Discrete Math. (to appear).

11. J. P. Hutchinson and G. L. Miller, "On deleting vertices to make a graph of positive genus planar," Discrete Algorithms and Complexity Theory, Proc. Japan-U.S. Joint Seminar, 1986, Kyoto, Perspectives in Computing, Vol. 15, Academic Press, Boston, 1987, 81-98.

12. A. Lempel, S. Even, and I. Cederbaum, "An algorithm for planarity testing of graphs," Theory of Graphs, Int'l. Symp., Rome, July, 1966, P. Rosenstiehl, ed., Gordon & Breach, N.Y., 1967, 215–232.

13. R. J. Lipton and R. E. Tarjan, "A separator theorem for planar graphs," SIAM J. Appl. Math., 36 (1979), 177–189.

14. C. Thomassen, "Embeddings of graphs with no short noncontractible cycles" (to appear).

15. A. T. White, Graphs, Groups and Surfaces, North-Holland, Amsterdam, 1973.

DEPARTMENT OF MATHEMATICS
SMITH COLLEGE
NORTHAMPTON, MASSACHUSETTS 01063

Contemporary Mathematics
Volume **89**, 1989

# INTERVAL HYPERGRAPHS

Arnold L. Rosenberg[1]

ABSTRACT. An $n$-vertex *interval hypergraph* (I-hypergraph, for short) $I$ comprises the set $V_n = \{1, 2, \cdots, n\}$ of vertices and a multiset $E(I)$ of hyperedges, each of the form $\{k, k+1, \cdots, k+r\}$ $(k \geq 1,\ 1 \leq r \leq n-k)$. One can decide in linear time whether or not a given hypergraph is an I-hypergraph. The *size* of $I$ is the sum of the cardinalities of its hyperedges. An *embedding* of the graph $G = (V, E)$ in $I$ comprises an injection $\mu_v : V \to V_n$, and an injection $\mu_e : E \to E(I)$, satisfying: for all $(u, v) \in E$, $\{\mu_v(u), \mu_v(v)\} \subseteq \mu_e(u, v)$. The problem of finding the smallest I-hypergraph in which a given graph can be embedded is *NP*-complete; so also is the problem of whether or not a given graph $G$ is embeddable in a given I-hypergraph $I$. Certain problems (e.g., 2-colorability and hamiltonianicity) that are *NP*-complete for arbitrary hypergraphs are solvable in polynomial time for I-hypergraphs. Say that the finite family of graphs $\Gamma$ has an $\alpha$-separator $(1/2 \leq \alpha < 1)$ of size $S(n)$. The $n$-vertex I-hypergraph $I$ is ***strongly universal*** for $\Gamma$ if: given any $W \subseteq V_n$ and any graph $G = (V, E) \in \Gamma$ with $|V| \leq |W|$, there is an embedding of $G$ in $I$ such that $\mu_v(V) \subseteq W$. There is an I-hypergraph of size

$$m \cdot \left( \sum_{k=1}^{\log m} \sum_{i=0}^{\lambda(2^k)} S(\alpha^i 2^k) \right),$$

that is strongly universal for $\Gamma$: $\lambda(M) =_{\text{def}} \log_{1/\alpha} M$; $m$ is the largest number of vertices in any $G \in \Gamma$. For the family of binary trees and any family for which $S(n)$ is of the form $n^\delta$, these strongly universal I-hypergraphs are within a constant factor of smallest possible.

[0]1980 *Mathematics Subject Classification* (1985 *Revision*). 94C15, 68C25, 05C99, 68E10.

[1]This research was supported in part by NSF Grants DCI-85-04308 and DCI-87-96236.

# 1. INTRODUCTION

This paper combines the research topics of two families of investigations that have appeared in the literature in recent years.

The first type of investigation expands the object of study in graph-embedding research from graphs to hypergraphs, motivated by the popularity of "bus-oriented" architectures in present-day microelectronics. Three examples are:

- In [4], Bhatt and Leiserson construct, for each integer $n$, what we are calling an interval-hypergraph in which every $n$-vertex binary tree can be embedded;
- in [14], Peterson and Ting determine (among other things) the minimum size of an interval-hypergraph in which the complete graph $K_n$ can be embedded;
- in [19], Stout studies mesh-structured processor arrays with busses as well as point-to-point communication links; he concludes (among other things) that "contiguous" busses are best from physical considerations.

The second type of investigation seeks, for a given finite family of graphs $\Gamma$, a graph $G(\Gamma)$ that is *strongly universal* for $\Gamma$ in the sense of containing each graph in $\Gamma$ as a subgraph, even if some positive fraction of the vertices of $G(\Gamma)$ are "killed".

- In [2], Beck establishes the existence of an $O(n)$-vertex $O(n)$-edge graph that remains universal for $(\leq n)$-vertex path graphs[2], even after some arbitrary positive fraction of the graph's vertices are killed;
- in [1], Alon and Chung present an explicit construction for the graphs advertised in [2];
- in [3], Beck studies the analogous problem for trees.
- in [7], Friedman and Pippenger show that expanding graphs contain all small trees, even after one kills a positive fraction of the vertices.

Some of the motivation for these studies is fault tolerance in arrays of processors.

In this paper, we are motivated by two issues concerning bus-oriented parallel architectures. First we wish to compare the properties of bus-oriented communication structures (as idealized by hypergraphs) as opposed to point-to-point communication structures (as idealized by graphs). Second, we wish to study the use of busses in designing fault-tolerant arrays of identical processors in an environment of VLSI (Very Large Scale Integrated) circuitry. To these ends, we formalize the notion of *interval hypergraph* studied informally in [4, 14, 16, 17], we study a number of the basic properties of interval hypergraphs, and we seek small interval hypergraphs that are *strongly universal* for given finite families of graphs, in the sense described above. The main result of our study is an algorithm that produces these small strongly universal interval hypergraphs. Included among the results we establish here are the following:

- The recognition problem for interval hypergraphs is solvable in linear time.
- The problem of finding the smallest interval hypergraph in which a given graph can be embedded is *NP*-complete.

[2]The $n$-vertex path-graph $P_n$ has vertices $\{1, 2, \cdots, n\}$ and edges $\{(i, i+1) \mid 1 \leq i < n\}$.

- The problem of deciding whether or not a given graph can be embedded in a given interval hypergraph is *NP*-complete.
- Certain problems, such as 2-colorability and hamiltonianicity that are *NP*-hard for general hypergraphs can be solved efficiently for interval hypergraphs.

We also make explicit the relationship between interval hypergraphs and the better-known interval graphs, both being characterized in terms of matrices with the so-called consecutive ones property.

Our main algorithm takes as input a finite family of graphs $\Gamma$ and an $\alpha$-separator function $S(n)$ for $\Gamma$ $(1/2 \leq \alpha < 1)$. The algorithm produces a strongly universal interval hypergraph $I(\Gamma)$ for $\Gamma$, of size (measured by the sum of the cardinalities of its hyperedges)[3]

$$SIZE(I(\Gamma)) = m \cdot \left( \sum_{k=1}^{\log m} \sum_{i=0}^{\lambda(2^k)} S(\alpha^i 2^k) \right),$$

where $m$ is the number of vertices in the largest graph in $\Gamma$, and $\lambda(M) =_{\text{def}} \log_{1/\alpha} M$. For many families $\Gamma$, including binary trees and any family for which $S(n)$ is of the form $n^\delta$, the interval hypergraphs $I(\Gamma)$ are optimal in *SIZE* (to within a constant factor). Moreover, for such $S(n)$, the *SIZE* of $I(\Gamma)$, which can be viewed as measuring the area required to lay $I(\Gamma)$ out in the plane, is just a small constant factor greater than the area of any "collinear" layout in the plane of the largest graph in $\Gamma$ ("collinearity" demanding that the graph's vertices lie along a line) .

# 2. BACKGROUND

## 2.1. The Formal Framework

We define the various notions that underlie the objects we study and the techniques we bring to bear on them.

*(Hyper)Graphs.* A *hypergraph* $H = (V, E)$ comprises a set $V$ of *vertices* and a multiset $E$ of subsets of $V$, called *hyperedges*. A *graph* $G = (V, E)$ is a hypergraph for which each $e \in E$ is a doubleton; each such hyperedge is called an *edge*. In this paper, we consider only graphs that are *simple* in the sense that $E$ is a *set*. All families of graphs considered here are finite.

*Interval Hypergraphs.* An $n$-vertex *interval hypergraph* (*I-hypergraph*, for short) $I$ is a hypergraph whose vertices comprise the set $V_n = \{1, 2, \cdots, n\}$ and whose hyperedges all have the form $\{k, k+1, \cdots, k+r\}$ for some $k \geq 1$ and $1 \leq r \leq n-k$.

Letting $G$ ambiguously denote a graph or a hypergraph, we denote by $|G|$ the number of vertices of the (hyper)graph $G$ and by *SIZE*$(G)$ the sum of the cardinalities of $G$'s (hyper)edges.

*Embedding.* An *embedding* of the graph $G = (V_g, E_g)$ in the hypergraph $H = (V_h, E_h)$ comprises one-to-one mappings

$$\mu_v : V_g \to V_h \quad \text{and} \quad \mu_e : E_g \to E_h$$

such that, for each edge $(u, v) \in E_g$, the image vertices $\mu_v(u)$ and $\mu_v(v)$ are both elements of the image hyperedge $\mu_e(u, v)$; symbolically, $\{\mu_v(u), \mu_v(v)\} \subseteq \mu_e(u, v)$. We say that a hypergraph *contains* any graph that is embeddable in it.

[3] All logarithms are to the base 2 unless indicated otherwise.

*Strong Universality.* Let $\Gamma$ be a finite family of graphs. The hypergraph $H = (V_h, E_h)$ is *strongly universal* for $\Gamma$ if, given any set $W \subseteq V_h$: for every graph $G = (V_g, E_g)$ in $\Gamma$ for which $|G| \leq |W|$,[4] there is an embedding of $G$ in $H$ with $\mu_v(V_g) \subseteq W$.

*Graph Separators and Separation Profiles.* Let $\alpha$ be a rational number in the range $1/2 \leq \alpha < 1$, and let $S(n)$ be a nondecreasing integer-valued function. The graph $G$ has an *$\alpha$-separator of size* $S(n)$ either if $|G| < 2$ or if the following holds: By removing at most $S(|G|)$ edges from $G$, one can partition $G$ into subgraphs $G_1$ and $G_2$, each of size

$$\lfloor (1-\alpha)|G| \rfloor \;\leq\; |G_i| \;\leq\; \lceil \alpha |G| \rceil,$$

and each having an $\alpha$-separator of size $S(n)$. A family of graphs $\Gamma$ has an $\alpha$-separator of size $S(n)$ iff each graph $G \in \Gamma$ does.

It is easy to verify that the family of path-graphs has a $(1/2)$-separator $S_{path}(n) \equiv 1$; Valiant [20] has shown that the family of binary trees has a $(2/3)$-separator of size $S_{tree}(n) \equiv 1$; it is an immediate consequence of the Planar Separator Theorem [11] that the family of rectangular meshes has a $(2/3)$-separator of size $S_{grid}(n) = \sqrt{8n}$. We use $\alpha$-separators here, rather than the more commonly used bisectors, since for certain families of graphs (e.g., binary trees), choices of $\alpha$ other than $\alpha = 1/2$ lead to strongly universal I-hypergraphs that are more *SIZE*-efficient by a logarithmic factor.

Let $G$ be a graph, and let $l$ be any integer $\geq \log_{1/\alpha} |G|$. The graph $G$ has an *$\alpha$-separation profile* ($\alpha$-*SP*, for short)

$$\langle s_l, s_{l-1}, \cdots, s_1 \rangle,$$

each $s_i$ a nonnegative integer, precisely if: by removing at most $s_l$ edges from $G$, one can partition $G$ into subgraphs $G_1$ and $G_2$, each of size $\leq \alpha|G|$, and each having an $\alpha$-SP

$$\langle s_{l-1}, s_{l-2}, \cdots, s_1 \rangle.$$

Another view of separation profiles is given by the notion of a

$$\langle s_l, s_{l-1}, \cdots, s_1 \rangle\text{-}\textit{decomposition tree}$$

for $G$: If one has a graph $G$ with an $\alpha$-SP

$$\langle s_l, s_{l-1}, \cdots, s_1 \rangle,$$

then one can construct a depth-$l$ binary tree whose root is $G$, and whose left and right subtrees are, respectively, the $\langle s_{l-1}, s_{l-2}, \cdots, s_1 \rangle$-decomposition trees of the graphs $G_1$ and $G_2$ mentioned above.

The notions "separator" and "separation profile" converge in the fact that every graph $G$ having an $\alpha$-separator of size $S(n)$ admits an $\alpha$-SP

$$\langle s_l, s_{l-1}, \cdots, s_1 \rangle,$$

where each $s_i = S(\alpha^{l-i}|G|)$. We leave to the reader the task of translating this correspondence into a decomposition tree for $G$. By dint of this relationship, we may refer freely to the *$S(n)$-decomposition tree* of any graph having an $\alpha$-separator of size $S(n)$.

---

[4] We denote by $|S|$ the cardinality of the set $S$.

### 2.2. The Intended Interpretation

In our motivating scenario, the graph $G$ represents a ***logical*** array: its vertices represent the processors of the array, and its edges represent communication links interconnecting the processors. (Thus, in $G$, interprocessor communication is "point-to-point"). The I-hypergraph $I$ represents a ***physical*** array we shall use to realize $G$: its vertices represent the processors of the array, and its hyperedges represent busses that the processors tap into, in order to realize the edges of the array. (Thus, in $I$, interprocessor communication is along busses.) A processor (vertex) can tap into any bus (hyperedge) it belongs to. $SIZE(I)$ approximates the area required to lay $I$ out in the plane (on a chip), using the following groundrules. The vertices of $I$ get laid out in a row, in natural order; the hyperedges get run as busses above the row, with vertical wires connecting each processor/vertex to the hyperedges it belongs to. Busses and wires have unit width; vertices occupy side-$s$ squares, where $s$ is large enough for the vertex to have its complement of incident edges. Busses and wires are allowed to cross – at most two crossing at a point – but not to overlap in any other way.

With the above representation in mind, the mapping $\mu_v$ can be viewed as assigning logical processors to physical processors, while $\mu_e$ assigns communication links to the busses that will realize them. The compatibility condition assures that any pair of processors that are supposed to use a bus can both be connected to it; the one-to-one condition assures that a hyperedge is used to realize at most one edge, modelling our assumption that each bus is dedicated to a single link.

When discussing fault tolerance, the set $W \subseteq V_h$ are the operational processors, while those in $W - V_h$ have failed; one wants to realize the array $G$ on the good processors of $I$.

### 2.3. Related Work

Aside from the motivating sources cited earlier, the research in this paper builds on the studies of I-hypergraphs in the following papers. In [4], one finds a construction of an $n$-vertex I-hypergraph in which one can embed any $n$-node binary tree. In [14], one finds lower bounds of the $SIZE$ of an I-hypergraph that contains the complete graph $K_n$, even when the notion of embedding is generalized so that the mapping $\mu_e$ need not be one-to-one (one uses time-sharing to reconcile contention for the busses). In [19], families of I-hypergraphs are presented, that are optimal in the sense of being able to simulate all other hypergraphs efficiently. In [16, 17, 6], our strong universality problem was first enunciated; $SIZE$-optimal strongly universal I-hypergraphs were presented for any finite family of path-graphs or of binary trees. Indeed, Section 4 of the present paper extends to families of arbitrary graphs the ideas and techniques of the last three cited papers.

## 3. BASIC PROPERTIES OF I-HYPERGRAPHS

We present in this section a number of basic properties of I-hypergraphs. Either the properties or their proofs will establish connections between I-hypergraphs and other structures that are better known.

### 3.1. Recognizing I-Hypergraphs

Our first result considers the problem of recognizing when a given hypergraph is an I-hypergraph. The proof of the result indicates an indirect connection between I-hypergraphs and interval graphs (i.e., intersection graphs of finite intervals on the real line), since both can be characterized in terms of an incidence matrix with the *consecutive ones* property.

**Proposition 1** *Given a hypergraph $H$, one can decide in time proportional to*

$$|H| + SIZE(H)$$

*whether or not $H$ is (isomorphic to) an interval hypergraph.*

*Proof.* Let $H$ have $h$ hyperedges. Consider the $|H| \times h$ (0,1)-valued *incidence matrix* for $H$, whose rows represent the vertices of $H$, whose columns represent the hyperedges of $H$, and whose $(i, j)$-th entry is 1 just when vertex $i$ belongs to hyperedge $j$. One verifies easily that $H$ is isomorphic to an interval hypergraph if, and only if, the rows of its incidence matrix can be permuted in such a way that all of the 1's in each column are consecutive. Booth and Lueker [5] present an algorithm that tests a (0,1)-valued matrix for this so-called *consecutive ones* property in time proportional to the sum of the number of rows (which is $|H|$) and the number of 1's (which is $SIZE(H)$), when the matrix is presented as a list of columns, with each column presented by a list of its 1-entries. □

## 3.2. Interval Graphs and Interval Hypergraphs

The proof of Proposition 1 has introduced the basic tool needed to expose the relationship between interval hypergraphs and their better-known relatives, interval graphs.

An *interval graph* is a graph $G$ whose vertices can be put in one-to-one correspondence with intervals of the real line in such a way that vertices $u$ and $v$ are adjacent in $G$ just when their corresponding intervals intersect.

Fulkerson and Gross [8] present a characterization of interval graphs in terms of a class of (0,1)-valued matrices. A *clique* in a graph $G$ is a maximal set of mutually adjacent vertices. Given a graph $G$ with $c$ cliques, define the *clique vs. vertex incidence matrix* of $G$ to be the $c \times |G|$ (0,1)-valued matrix whose rows represent the cliques of $G$, whose columns represent the vertices of $G$, and whose $(i, j)$-th entry is 1 just when vertex $j$ belongs to clique $i$. Fulkerson and Gross establish the following.

**Lemma 1** [8] *A graph is an interval graph if, and only if, its clique vs. vertex incidence matrix has the consecutive ones property.*

Thus, in some sense, the vertices of interval hypergraphs correspond to the cliques of interval graphs, while the hyperedges of interval hypergraphs correspond to the vertices of interval graphs. This correspondence can be made tight in one direction. Given a graph $G$, construct the hypergraph $H(G)$ as follows.

- For each clique $\kappa$ of $G$, add a unique vertex $v(\kappa)$ to $H(G)$; every vertex of $H(G)$ arises in this way.
- For each vertex $v$ of $G$, add a unique hyperedge to $H(G)$ containing every vertex $v(\kappa)$ of $H(G)$ for which the clique $\kappa$ of $G$ contains vertex $v$ of $G$.

**Proposition 2** *For any graph $G$, the clique vs. vertex incidence matrix of $G$ is identical to the incidence matrix of the hypergraph $H(G)$. Hence, in particular, $G$ is an interval graph if, and only if, $H(G)$ is an interval hypergraph.*

Thus, every interval hypergraph can be viewed as coming from an interval graph, and every interval graph spawns an interval hypergraph.

One can prove a weak converse of Proposition 2, showing how certain interval hypergraphs spawn interval graphs.

Given a hypergraph $H$, construct the graph $G(H)$ as follows.

- For each hyperedge $\eta$ of $H$, there is a unique vertex $v(\eta)$ of $G(H)$; every vertex of $G(H)$ arises in this way.
- Two vertices $v(\eta_1)$ and $v(\eta_2)$ of $G(H)$ are adjacent just when the hyperedges $\eta_1$ and $\eta_2$ of $H$ have nonempty intersection.

Say that the hypergraph $H$ has the *vertex isolation property* if every vertex of $H$ is the unique common element of some subset of the hyperedges of $H$.

**Proposition 3** *For any hypergraph $H$ having the vertex isolation property, the incidence matrix of $H$ is identical to the clique vs. vertex incidence matrix of $G(H)$. Hence, in particular, $G(H)$ is an interval graph if, and only if, $H$ is an interval hypergraph.*

The straightforward proofs of Propositions 2 and 3 are left to the reader.

## 3.3. Finding Small I-Hypergraphs

If the hypergraph $H$ is not (isomorphic to) an I-hypergraph, one might wish to find the smallest I-hypergraph that contains $H$. Our next result indicates that finding this small I-hypergraph is likely to be computationally intractable, even when $H$ is a graph. This demonstration exposes a connection between I-hypergraphs and the *Optimal Linear Arrangement Problem* for graphs.

**Proposition 4** *The following problem is* NP*-complete: Given a graph $G$ and an integer $S$, to decide whether or not there exists an I-hypergraph of SIZE $S$ that contains $G$. The problem is solvable in polynomial time when $G$ is a tree.*

*Proof.* The result will follow from a demonstration that our *Smallest Containing I-Hypergraph Problem* (**SCIH**) is *equivalent* to the Optimal Linear Arrangement Problem for graphs [10] (**OLA**), in the sense that each problem is reducible to the other in polynomial time[5]. The reducibility of **OLA** to **SCIH** establishes the result for arbitrary graphs [10]; the reducibility of **SCIH** to **OLA** establishes the result for trees [18]. **OLA** is defined as follows.

**OLA**: Given a graph $G = (V, E)$ and an integer $B$, to decide whether or not there exists an injection $\lambda : V \to \{1, 2, \cdots, |G|\}$ for which

$$\sum_{(u,v)\in E} |\lambda(u) - \lambda(v)| \leq B.$$

The bases for our claimed reducibilities are the following correspondences. For any graph $G = (V, E)$ and injection $\lambda : V \to \{1, 2, \cdots, |G|\}$, $G$ is embeddable, via the vertex-injection $\mu_v = \lambda$, in the I-hypergraph $I_\lambda$ that has vertex-set $\{1, 2, \cdots, |G|\}$ and that has a hyperedge

$$\{\lambda(u), \lambda(u) + 1, \cdots, \lambda(v)\}$$

[5] Since all of the reductions we present in this section are "in polynomial time", we shall henceforth leave the phrase to be understood implicitly.

for each edge $(u, v) \in E$; in the embedding, $\mu_e$ maps the edge $(u, v)$ to this hyperedge. The hypergraph $I_\lambda$ is easily seen to have *SIZE*

$$|E| + \sum_{(u,v)\in E} |\lambda(u) - \lambda(v)|.$$

Conversely, let the graph $G = (V, E)$ be embedded in the I-hypergraph $I = (V_h, E_h)$ via the injections $\mu_v$ and $\mu_e$. Then the injection $\lambda = \mu_v$ has

$$\sum_{(u,v)\in E} |\lambda(u) - \lambda(v)| \le SIZE(I) - |E|,$$

since for each edge $(u, v)$, $|\lambda(u) - \lambda(v)| \le |\mu_e(u, v)| - 1$. The constructions outlined here can obviously be effected in polynomial time.

These correspondences demonstrate that the graph $G = (V, E)$ admits a linearization with **OLA**-cost $B$ if, and only if, $G$ is embeddable in an I-hypergraph of size $B + |E|$. It follows that the problems **OLA** and **SCIH** can each be efficiently reduced one to the other. □

### 3.4. Deciding Embeddability

Lacking the ability to determine efficiently how small an I-hypergraph the graph $G$ can be embedded in, one might want at least to determine if $G$ can be embedded in a given I-hypergraph $I$. We now show that this problem, too, is likely to be computationally intractable. This result illustrates a connection between I-hypergraphs and the *Bandwidth Minimization Problem* for graphs.

**Proposition 5** *The following problem is* NP*-complete: Given a graph $G$ and an I-hypergraph $I$, to decide whether or not there exists an embedding of $G$ in $I$. The problem remains* NP*-complete even when $G$ is a binary tree.*

*Proof.* The result will follow from a demonstration that the Bandwidth Minimization Problem for graphs (**BMP**) is reducible to our *I-Hypergraph Embeddability Problem* (**IHEP**). This reducibility establishes the result for arbitrary graphs because of [13], and for trees because of [9]. **BMP** is defined as follows.

**BMP**: Given a graph $G = (V, E)$ and an integer $B < |G|$,[6] to decide whether or not there exists an injection $\lambda : V \to \{1, 2, \cdots, |G|\}$ for which

$$\max_{(u,v)\in E} |\lambda(u) - \lambda(v)| \le B.$$

The base for our claimed reducibility is the following correspondence. For any graph $G = (V, E)$ and integer $B$, construct the I-hypergraph $I_{G,B}$ that has vertex-set $\{1, 2, \cdots, |G|\}$ and that has, for each $1 \le i \le |G| - B + 1$, $\min\left(\binom{B}{2}, |E|\right)$ copies of the hyperedge

$$\{i, i + 1, \cdots, i + B - 1\}.$$

One verifies easily that the hypergraph $I_{G,B}$ can be constructed in time polynomial in the size of the description of $G$ and $B$, since

$$SIZE(I_{G,B}) \le |G|^2 \cdot \min\left(\binom{B}{2}, |E|\right) < |G|^4.$$

[6]The inequality "$B < |G|$" is not usually included in the definition of **BMP**, but when it does not hold, the decision problem trivializes.

To establish the reduction, assume first that the graph $G$ admits a layout $\lambda$ for which

$$\max_{(u,v)\in E} |\lambda(u) - \lambda(v)| \leq B.$$

Then $G$ is embeddable in $I_{G,B}$: The vertex-injection is $\mu_v = \lambda$; the edge-injection is defined by: $\mu_e(u, v)$ is an arbitrary one of the hyperedges of the form

$$\{\lambda(u), \lambda(u) + 1, \cdots, \lambda(u) + B - 1\}.$$

(We have endowed $I_{G,B}$ with ample copies of the hyperedge to guarantee that we can construct the injection $\mu_v$.) Conversely, say that $G$ is embeddable in $I_{G,B}$ via the injections $\mu_v$ and $\mu_e$. Then the layout $\lambda = \mu_v$ has

$$|\lambda(u) - \lambda(v)| \leq |\mu_e(u, v)| = B$$

for each edge $(u, v)$ of $G$, so $G$ has bandwidth no greater than $B$. The result follows. □

## 3.5. Deciding 2-Colorability

A hypergraph $H$ is *2-colorable* if there is a way to assign one of two distinct colors to each vertex in $H$ in such a way that no hyperedge of $H$ remains monochromatic. Lovasz [12] has shown that the problem of deciding, given an arbitrary hypergraph $H$, whether or not $H$ is 2-colorable is $NP$-complete. When $H$ is a graph, the same question is easily decided in time $O(|V| + |E|)$, for a graph is 2-colorable if and only if it is bipartite. Deciding 2-colorability of I-hypergraphs is even easier, as the following observation verifies.

**Proposition 6** *Every I-hypergraph is 2-colorable.*

*Proof.* Since every hyperedge of an I-hypergraph contains at least two vertices, and since the vertices of each hyperedge are "contiguous", one can 2-color an I-hypergraph by assigning one color to its odd-numbered vertices and another color to its even-numbered vertices. □

## 3.6. Deciding Connectivity and Path-Connectivity

An $n$-vertex hypergraph $H$ is *connected* (resp., *path-connected*) if $H$ contains an $n$-vertex tree (resp., the $n$-vertex path-graph $P_n$). The problem of deciding whether or not a given general hypergraph is path-connected is $NP$-complete, since it subsumes the problem of testing a graph for the existence of a hamiltonian path. In contrast, one can test a given I-hypergraph for path-connectivity via a straightforward efficient algorithm, because an I-hypergraph contains a spanning path-graph if, and only if, the path-graph is embeddable in the "natural" way. Moreover, again in contrast to arbitrary hypergraphs (or graphs, for that matter), an I-hypergraph is connected if, and only if, it is path-connected. We approach our decision algorithm via the following chain of lemmas.

**Lemma 2** *Let the $n$-vertex tree $T$ be embedded in the plane, with its vertices in a row, in the order*

$$v_1, v_2, \cdots, v_n.$$

*Then: for each $1 \leq k < n$, at least $k$ edges of $T$ have one vertex in the set $\{v_1, v_2, \cdots, v_k\}$; at least one of these edges must also have a vertex in the set $\{v_{k+1}, v_{k+2}, \cdots, v_n\}$. Perforce, the same is true for the path-graph $P_n$, since it is an $n$-vertex tree.*

*Proof.* Let us count the edges of $T$ that contain[7] a vertex in the set $V_k =_{\text{def}} \{v_1, v_2, \cdots, v_k\}$. Say that $V_k$ contains $l$ connected components $C_1, C_2, \ldots, C_l$ of $T$, of sizes $s_1, s_2, \ldots, s_l$, respectively. By dint of its being connected, each component $C_i$ must contain both vertices of at least $s_i - 1$ edges of $T$. Since $T$ is connected, each of these components must also contain at least one vertex belonging to an edge whose other vertex resides in the set $\{v_{k+1}, v_{k+2}, \cdots, v_n\}$. The result now follows by counting. □

**Lemma 3** *Let the n-vertex I-hypergraph $I$ contain an n-vertex tree $T$. Then, for each $1 \leq k < n$, at least $k$ hyperedges of $I$ have their smallest numbered vertex in the set $\{1, 2, \cdots, k\}$; at least one hyperedge has its highest numbered vertex in the set $\{k+1, k+2, \cdots, n\}$. Perforce, the same is true if $I$ contains the path-graph $P_n$, since it is an n-vertex tree.*

*Proof.* Any embedding of $T$ in $I$ induces a layout of $T$ in the plane with the vertices lying in a row; just let each $v_i = \mu_v^{-1}(i)$. Since each hyperedge of $I$ realizes just one edge of $T$, the result is immediate from Lemma 2. □

**Lemma 4** *If the n-vertex I-hypergraph $I$ contains an n-vertex tree (perforce, if it contains the path-graph $P_n$), then $I$ contains the path-graph $P_n$ embedded via the vertex-injection $\mu_v : V_{P_n} \rightarrow V_I$ defined by*

$$\mu_v(i) = i$$

*for $1 \leq i \leq n$.*

*Proof.* It follows by an induction based on Lemma 3 that the following greedy algorithm specifies a valid edge-injection $\mu_e : Edges(P_n) \rightarrow Hyperedges(I)$ to complement the injection $\mu_v$ defined in the statement of the Lemma.

Proceed along the row of vertices from left to right. For each vertex $i$, assign edge $(i, i+1)$ to any hyperedge of $I$ that

- contains vertices $i$ and $i+1$
- has not yet been used
- has minimal largest element among hyperedges that have not yet been used

Details are left to the reader. □

**Proposition 7** *Given an I-hypergraph $I$, one can determine in time*[8] $O(\mathcal{H}(SIZE(I)))$ *whether or not $I$ is connected or, equivalently, path-connected.*

*Proof.* Let $I$ have $h$ hyperedges. Say that $I$ is presented via the $h \times 2$ matrix $M_I$ that associates with each hyperedge of $I$ its minimum and its maximum element: this matrix is clearly obtained from the incidence matrix of $I$ in time linear in $SIZE(I)$. Reorder the columns of $M_I$ so that the hyperedges of $I$ are ordered by increasing minimum element and, among hyperedges with the same minimum element, by increasing maximum element. This reordering requires at most time $O(\mathcal{H}(h))$, which is obviously $O(\mathcal{H}(SIZE(I)))$. Once $M_I$ is so rearranged, it is a simple matter to implement the algorithm of Lemma 4 in time linear in $SIZE(I)$. □

[7] This terminology is justified by our defining an edge as a two-element set.

[8] For all $s$, $\mathcal{H}(s) = s \cdot \log s$.

### 3.7. Finding a "Good" I-Hypergraph

Finally, we indicate how to construct a small I-hypergraph that contains a given graph $G$, based on a separator for $G$. This construction yields one more connection between the problem of embedding graphs in I-hypergraphs and the general problem of finding collinear layouts of graphs.

**Proposition 8** *Let the graph $G$ have an $\alpha$-separator of size $S(n)$ for some $1/2 \leq \alpha < 1$. Then $G$ is embeddable in an I-hypergraph $I(G)$ of SIZE at most*

$$|G| \cdot \left( \sum_{i=0}^{\lambda(|G|)} S(\alpha^i |G|) \right),$$

*where* $\lambda(|G|) = \log_{1/\alpha}(|G|)$.

*Proof.* We employ a strategy derived from [15]. Given a graph $G$, we construct $I(G)$ and the embedding-injections $\mu_v$ and $\mu_e$ as follows.

The I-hypergraph $I(G)$ has vertex-set $\{1, 2, \cdots, |G|\}$. To specify the injection $\mu_v$, construct an $S(n)$-decomposition tree for $G$, as described in Section 2. Place the vertices of $G$ in a row in the order they occur as leaves of the decomposition tree. This ordering implicitly specifies $\mu_v$; it also implicitly lays out, in contiguous blocks, the vertices of all of the subgraphs of $G$ that occur in the decomposition tree.

We now specify the hyperedges of $I(G)$ and the injection $\mu_e$. For each of the (at most $S(|G|)$) edges that interconnect the two subgraphs $G_1$ and $G_2$ of $G$, at level-1 of the decomposition tree, give $I(G)$ a hyperedge $\{1, 2, \cdots, |G|\}$; let $\mu_e$ associate each connecting edge with a unique one of these hyperedges. Next: For each of the (at most $S(\alpha|G|)$) edges that interconnect the subgraphs $G_{11}$ and $G_{12}$ of $G$, at level-2 of the decomposition tree, give $I(G)$ the hyperedge $\{1, 2, \cdots, |G_1|\}$, and let $\mu_e$ associate each connecting edge with a unique one of these hyperedges; similarly, for each of the (at most $S(\alpha|G|)$) edges that interconnect the subgraphs $G_{21}$ and $G_{22}$ of $G$, at level-2 of the decomposition tree, give $I(G)$ the hyperedge $\{|G_1| + 1, |G_1| + 2, \cdots, |G|\}$, and let $\mu_e$ associate each connecting edge with a unique one of these hyperedges. We continue in the indicated fashion to add hyperedges to $I(G)$ for "routing" the interconnections among the subgraphs of $G$ in the decomposition tree, using at most $S(\alpha^k|G|)$ copies of each hyperedge for the $2^{k-1}$ pairs of subgraphs at level-$k$ of the tree. Once having completed this construction, we shall have constructed $I(G)$ and embedded $G$ in it. It is clear from the construction that *SIZE*$(I(G))$ is bounded as claimed in the statement of the Proposition. □

## 4. STRONGLY UNIVERSAL I-HYPERGRAPHS

We turn now to the main result of this paper. Throughout this section, assume that we have been given the desired family of graphs $\Gamma$, where the largest graph in $\Gamma$ has $m$ vertices. For convenience, say that $m = 2^r$ is a power of 2. Let $\Gamma$ have an $\alpha$-separator of size $S(n)$ for some $1/2 \leq \alpha < 1$.

**Theorem 1** *Let the family of graphs $\Gamma$, as described above, be given. There is a strongly universal I-hypergraph $I(\Gamma)$ for $\Gamma$ of size*

$$SIZE(I(\Gamma)) = m \cdot \left( \sum_{k=1}^{r} \sum_{i=0}^{-k/\log \alpha} S(\alpha^i 2^k) \right). \tag{1}$$

The remainder of the section is devoted to proving Theorem 1, i.e., describing $I(\Gamma)$ and verifying that it is indeed strongly universal for $\Gamma$.

**Remark.** In order to reconcile Equation (1) with the expression in the Abstract, recall that $-k/\log\alpha = \log_{1/\alpha} 2^k$.

## 4.1. The Construction of $I(\Gamma)$ and the Embedding Procedure

Let the vertices of $I(\Gamma)$ be the set $V_m = \{1, 2, \cdots, m\}$. We give $I(\Gamma)$ the following hyperedges: for $k = 1, 2, \cdots, r$ and $a = 0, 1, 2, \cdots, 2^{r-k} - 1$, we create $\sum_{i=0}^{-k/\log\alpha} S(\alpha^i 2^k)$ copies of the hyperedge

$$\{a2^k + 1, a2^k + 2, \cdots, (a+1)2^k\}.$$

It is clear that the $I(\Gamma)$ just constructed has size

$$SIZE(I(\Gamma)) = m \cdot \left( \sum_{k=1}^{r} \sum_{i=0}^{-k/\log\alpha} S(\alpha^i 2^k) \right),$$

as claimed in the Theorem.

Although formal validation of $I(\Gamma)$ will wait for the next subsection, we indicate informally how the graphs in $\Gamma$ are embedded in arbitrary vertex-subsets of $I(\Gamma)$. Say that we are told that the $p$ vertices

$$v_1, v_2, \cdots, v_p$$

(each $v_j \in \{1, 2, \cdots, m\}$) of $I(\Gamma)$ are the available ones and that we are to realize the ($\leq p$)-vertex graph $G \in \Gamma$ on these vertices. We begin the embedding process by constructing a $S(n)$-decomposition tree for $G$. We then lay out the vertices of $G$ on the available vertices of $I(\Gamma)$, in the order in which the vertices occur as leaves of the $S(n)$-decomposition tree. (If $G$ has fewer than $p$ vertices, we arbitrarily choose $|G|$ of the available vertices for $G$'s vertices.) Thus we have the vertex-injection $\mu_v$. In order to specify the edge-injection $\mu_e$, we associate with edge $(u, v)$ of $G$ any as-yet unused smallest hyperedge of $I(\Gamma)$ that contains both $\mu_v(u)$ and $\mu_v(v)$.

## 4.2. The Construction Validated

We now validate the construction and embedding process of the previous subsection. Our validation uses a new graph-theoretic notion motivated by the stringent demands of strong universality.

*Strong Separation Profiles.* Our interval hypergraphs $I(\Gamma)$ decompose naturally by bisection. Removing the largest hyperedges decomposes $I(\Gamma)$ into two copies of the I-hypergraph that we would construct if all graphs of size $> m/2$ were removed from $\Gamma$, and so on. When a graph $G$ is embedded in $I(\Gamma)$, it is not clear how this bisection will dissect $G$, for that depends on which vertices of $I(\Gamma)$ are declared available for the embedding. Our guarantee that $G$ can be embedded no matter which vertices of $I(\Gamma)$ are available thus leads naturally to the following notion.

Assume throughout that $n$ (the number of vertices in $I(\Gamma)$) is a power of 2. Let $G$ be a graph with $n$ or fewer vertices, and let $l$ be any integer $\geq \log n$. The $l$-tuple of nonnegative integers

$$\langle e_l, e_{l-1}, \cdots, e_1 \rangle$$

is a *strong separation profile* (an *SSP*, for short) for $G$, if the following property holds.

*The SSP Property:* Given any integer $n_1$ such that both $n_1$ and $|G| - n_1$ are $\leq n/2$: By removing at most $e_l$ edges from $G$, one can partition $G$ into subgraphs $G_1$ having $n_1$ vertices and $G_2$ having $|G| - n_1$ vertices, each of which has $\langle e_{l-1}, e_{l-2}, \cdots, e_1 \rangle$ as an SSP. This recursive decomposition of $G$ continues until we get down to subgraphs of $G$ having at most one vertex.

Note that one can view each candidate decomposition of $G$ (corresponding to the different choices for $n_1$) in terms of an $\langle e_l, e_{l-1}, \cdots, e_1 \rangle$-*decomposition tree* for $G$: the tree's root is $G$, with sons $G_1$ and $G_2$, and so on, just as with the $S(n)$-decomposition trees of the earlier sections.

The "strong" in the term SSP is intended to contrast with the notion of $\alpha$-SP, wherein one seeks a "small cut" partition for just the case $\lfloor (1-\alpha)|G| \rfloor \leq n_1 \leq \lceil \alpha |G| \rceil$, rather than for all values of $n_1$, $1 \leq n_1 \leq n/2$.

The relevance of the notion of SSP resides in the following result.

**Lemma 5** *Given any $l$-tuple of nonnegative integers*

$$\tau = \langle e_l, e_{l-1}, \cdots, e_1 \rangle$$

*one can construct an $(m = 2^l)$-vertex I-hypergraph $I(m)$ of size*

$$SIZE(I(m)) = m \cdot \sum_{i=1}^{l} e_i$$

*that is strongly universal for the family $\Gamma(\tau)$, where $\Gamma(\tau)$ comprises all graphs having the tuple $\tau$ as an SSP.*

*Proof.*

*The I-Hypergraph $I(m)$*

To construct $I(m)$, we create the following hyperedges from the vertex-set $V_m = \{1, 2, \cdots, m\}$. For $k = 1, \cdots, l$ and $a = 0, 1, 2, \cdots, 2^{l-k} - 1$, we create $e_k$ copies of the hyperedge

$$\{a2^k + 1, a2^k + 2, \cdots, (a+1)2^k\}.$$

It is clear that $I(m)$, so constructed, has the claimed *SIZE*.

*The Embedding Procedure*

Say that we are told that the $p$ vertices

$$v_1, v_2, \cdots, v_p$$

(each $v_j \in \{1, 2, \cdots, m\}$) of $I(m)$ are available and that we are to embed the $(\leq p)$-vertex graph $G \in \Gamma(\tau)$ on these vertices. The essence of the embedding process is the construction of an $\langle e_l, e_{l-1}, \cdots, e_1 \rangle$-decomposition tree for $G$. We begin by choosing some $|G|$ of the available vertices of $I(m)$ upon which to place the vertices of $G$; these vertices can be chosen *in any way whatsoever*. This choice then determines the parameter $n_1$, which is the size of one of the two graphs we shall partition $G$ into: Specifically,

$$n_1 =_{\text{def}} |\{v_j : v_j \leq 2^{l-1}\}|;$$

i.e., $n_1$ is the number of selected available vertices that reside to the left of the midpoint $m/2$ of $I(m)$. By definition of SSP, $G$ can be partitioned into a subgraph of size $n_1$ and one of size $|G| - n_1$ by removing no more than $e_l$ edges from $G$. These edges can thus be embedded in the $e_l$ size-$m$ hyperedges of $I(m)$, no matter which vertices of $I(m)$ their endpoints are placed on. By definition of SSP, we may assume that each of the two resulting subgraphs has an SSP

$$\langle e_{l-1}, e_{l-2}, \cdots, e_1 \rangle.$$

We thus find ourselves with two half-size versions of our original problem: By removing the $e_l$ large hyperedges from $I(m)$, we are left with two copies of $I(m/2)$ in which to embed the two subgraphs of $G$, each by definition having no more than $2^{l-1}$ vertices. We leave to the reader the easy details of inductively validating this recursive embedding process (which can be viewed as building an $\langle e_l, e_{l-1}, \cdots, e_1\rangle$-decomposition tree for $G$). □

Determining SSPs for arbitrary graphs is not a trivial pursuit. However, one can, with little difficulty, discover profiles for certain familiar graphs. For instance, every ($\leq n$)-vertex binary tree has an SSP of the form

$$\langle \log n, \log(n/2), \cdots, 1\rangle$$

so $e_{k-1} = e_k - 1$; similarly, every ($\leq n$)-vertex rectangular mesh has an SSP of the form

$$\langle \sqrt{n}, \sqrt{n}/\sqrt{2}, \sqrt{n}/2, \cdots, 1\rangle$$

so $e_{k-1} = e_k/\sqrt{2}$.[9] The following Lemma helps one discover SSPs; and it combines with Lemma 5 to complete the proof of Theorem 1.

**Lemma 6** *Let $\Gamma$ be a family of graphs having an $\alpha$-separator of size $S(n)$. For every integer $r$, every graph $G \in \Gamma$ with $|G| \leq 2^r$ has an SSP*

$$\langle e_r, e_{r-1}, \cdots, e_1\rangle,$$

*where each*

$$e_k = \sum_{i=0}^{-k/\log\alpha} S(\alpha^i 2^k).$$

*Proof.* The proof builds on the technique used in the proof of Proposition 8 for laying out (within the groundrules of Section 2.2) any given $G \in \Gamma$; therefore, we shall be very sketchy here. Note that the layout here (in contrast to that in Proposition 8) is purely a technical device and should not be construed as an embedding of $G$ in an I-hypergraph, despite the formal similarity.

Construct an $S(n)$-decomposition tree for $G$, and place the vertices of $G$ in a row in the order they occur as leaves of the decomposition tree. Run $S(|G|)$ routing tracks above the vertices, in which to route the edges that interconnect the two subgraphs $G_1$ and $G_2$ of $G$ at level-1 of the decomposition tree. These routing tracks can be viewed as rows in the plane that are reserved for "drawing" the edges of $G$; thus every edge of $G$ ends up being drawn as two vertical line segments from its terminal vertices to the associated routing track, plus a horizontal line segment (in the routing track) joining the two vertical segments. Then run $S(\alpha|G|)$ routing tracks over the vertices of $G_1$ and the same number of routing tracks over the vertices of $G_2$. Continue in the indicated fashion to run routing tracks for routing the edges among the subgraphs of $G$ in the decomposition tree, using $S(\alpha^k|G|)$ routing tracks for the $2^{k-1}$ pairs of subgraphs at level-$k$ of the tree. The reader will note that we have constructed here a layout of $G$ that uniformly has

$$W = \sum_{i=0}^{\log_{1/\alpha}|G|} S(\alpha^i|G|)$$

routing tracks above every vertex. It follows that, given any integer $n \leq |G|$, $G$ can be partitioned into a subgraph of size $n$ and one of size $|G| - n$ by removing (or "cutting") at most $W$ edges. In particular, such a partition is possible for any $n$ such that both $n$ and $|G| - n$ are $\leq 2^{r-1}$. □

Lemmas 5 and 6 combine to establish Theorem 1.

[9]The cited SSPs for trees and meshes can be derived by considering the sizes of "perimeters" of regions within the graphs.

## 4.3. The Issue of Optimality

There are many families of graphs for which our strongly universal I-hypergraphs are within a constant factor of optimal in *SIZE*. We cite two major examples.

Let us restrict attention to *honest* separation functions for families $\Gamma$, i.e., separator functions $S(n)$ that truly reflect the difficulty of cutting the member graphs into pieces, in the sense that $\Theta(S(n))$ edges are necessary, as well as sufficient, to partition an $n$-vertex graph in $\Gamma$ into two subgraphs of appropriate sizes.

*Binary Trees.* It is shown in [6] that any I-hypergraph that is strongly universal for the family of binary trees (which admits the honest $(2/3)$-separator function $S(n) \equiv 1$) has *SIZE* $\Omega(n \cdot \log^2 n)$, which is within a constant factor of the *SIZE* of the I-hypergraph produced by the construction in the proof of Theorem 1.

*Algebraic Separators.* Let the family $\Gamma$ have an honest $\alpha$-separator of size $S(n) = n^\delta$ for some constant $\delta$. The double summation (1) in Theorem 1 becomes a double geometric sum, so the construction in the proof of the Theorem yields an I-hypergraph of *SIZE* $O(n^{1+\delta})$. On the other hand, invoking the honesty of $S(n)$, we can invoke the bounding techniques of [15] to show that any I-hypergraph that is strongly universal for $\Gamma$ must have *SIZE* $\Omega(n^{1+\delta})$; indeed any collinear layout of the graphs in $\Gamma$ must occupy this much area, so in this case, there is at most constant factor overhead for the fault tolerance afforded by the strong universality of the I-hypergraphs we produce.

## 4.4. A Remaining Challenge

In [17], we studied the strong universality problem for the family $\Pi_n$ of path-graphs containing at most $n$ vertices. We were able to show there that one could sometimes produce "strongly universal" I-hypergraphs of smaller *SIZE*, if one moderated one's demands on strong universality so that one was guaranteed to be able to embed a given $G$ from the target graph family $\Gamma$ ($\Pi_n$ in that paper) *only with very high probability.* Specifically, the following results appear in that paper.

**Proposition 9** [17]
(a) *For all $n$, there exists an $n$-vertex I-hypergraph $I_n$ of SIZE $O(n \cdot \log n)$ that is strongly universal for the family $\Pi_n$.*
(b) *Any $n$-vertex I-hypergraph that is strongly universal for the family $\Pi_n$ must have SIZE $\Omega(n \cdot \log n)$.*

Now, let us change the game somewhat by assuming that when we "kill" vertices of the I-hypergraphs in question (i.e., make them unavailable), we do so independently, with probability 1/2. We can now consider the situation where an I-hypergraph is strongly universal with some given probability (which depends on the probabilities of certain patterns of "kills"). The following result concerns such a scenario.

**Proposition 10** [17]
(a) *For all $n$, there exists an $n$-vertex I-hypergraph $J_n$ of SIZE $O(n \cdot \log\log n)$ that is strongly universal for the family $\Pi_n$, with probability*

$$1 - \frac{1}{2n \cdot \log n}.$$

(b) *For $n > 4$, any $n$-vertex I-hypergraph that is strongly universal for the family $\Pi_n$ with probability at least*

$$1 - \frac{1}{2n \cdot \log n}$$

*must have SIZE $\Omega(n \cdot \log\log n)$.*

In order to lend intuition to the reader, we sketch the proofs of these results very briefly.

The nonprobabilistic upper bound of Proposition 9(a) proceeds much as in the proof of Theorem 1 in Section 4.1: Assume for simplicity that $n$ is a power of 2. For $k = 1, \cdots, l$ and $a = 0, 1, 2, \cdots, 2^{l-k} - 1$, we endow $I_n$ with one copy of the hyperedge

$$\{a2^k + 1, a2^k + 2, \cdots, (a+1)2^k\}.$$

The proof that $I_n$ is strongly universal for $\Pi_n$ consists of showing that one can always embed any sufficiently small path-graph by associating vertices of the path-graph with available vertices of $I_n$ in any order-preserving way, and realizing each edge of the path-graph via the smallest hyperedge that contains the images under $\mu_v$ of the edge's endpoints.

The nonprobabilistic lower bound of Proposition 9(b) proceeds by noting that, given any $n$-vertex I-hypergraph $I$ that is strongly universal for the family $\Pi_n$, we must be able to embed the $2^{k+1}$-vertex path-graph in $I$, using vertices

$$1, n/2^k, n/2^k + 1, \cdots, n - n/2^k, n - n/2^k + 1, n$$

of $I$, for each $k \in \{0, 1, \cdots, \log n\}$. Each value of $k$ thus contributes roughly $n$ to *SIZE*$(I)$, for a total of $\Omega(n \cdot \log n)$. (One must take some care in counting these contributions to *SIZE*$(I)$, since hyperedges added to satisfy the requirements of a small value of $k$ can be reused in satisfying the requirements of larger values of $k$.)

The probabilistic upper bound of Proposition 10(a) follows from a composite construction of $J_n$. Assume as before that $n$ is a power of 2. Partition the set $\{1, 2, \cdots, n\}$ into contiguous blocks of length $m = 2\log n$ each. Assign hyperedges to each block to make it a copy of $I_m$, as described above. Additionally, for each pair of adjacent blocks, add a hyperedge that is the union of the vertices in the two blocks. One now verifies that given any selection of available vertices of $J_n$, one can embed a path-graph using all of those vertices *unless* there are two available vertices separated by a block containing no available vertex. However, the probability of such an occurrence is no greater than $1/(2n \cdot \log n)$.

Finally, to see the probabilistic lower bound of Proposition 10(b), partition the vertices of any given $n$-vertex I-hypergraph $I$ into contiguous blocks of $m = \log n + \log\log n + 1$ vertices each. Assume that for each block $B$, there is some pattern of "killed" vertices that makes it impossible to embed a path-graph on all of the available vertices. The probability that $I$ can embed any sufficiently small path-graph cannot exceed the probability that one of these bad patterns occurs, which the reader can easily show to be greater than $1/(2n \cdot \log n)$. As a consequence, one can show that if $I$ successfully embeds path graphs with probability exceeding $1 - 1/(2n \cdot \log n)$, then it must *always* work on every one of the blocks $B$, hence must have *SIZE* at least as great as $I_m$ (by Proposition 9(b)).

Details on all four proofs are found in [17].

*The Challenge:* We are certain that savings analogous to those exposed in Propositions 9 and 10 are attainable with strongly universal I-hypergraphs for a large variety of graph families other than $\Pi_n$, but we have as yet been unable to generalize this phenomenon even to binary trees. Such generalization is an inviting challenge.

**ACKNOWLEDGMENT.** It is a pleasure to thank Lenny Heath and Bruce Leban for a careful reading of the manuscript.

# 5. REFERENCES


1. N. Alon and F.R.K. Chung, "Explicit constructions of linear-sized fault-tolerant networks," Typescript, MIT (1985).

2. J. Beck, "On size Ramsey number of paths, trees, and circuits, I," *J. Graph Th.*, *7* (1983) 115-129.

3. J. Beck, "On size Ramsey number of paths, trees, and circuits, II," Manuscript (1983).

4. S.N. Bhatt and C.E. Leiserson, "How to assemble tree machines," pp. 95-114 in *Advances in Computing Research 2*, (F.P. Preparata, ed.) JAI Press, Greenwich, CT, 1984.

5. K.S. Booth and G.S. Lueker "Testing for the consecutive ones property, interval graphs, and graph planarity using PQ-tree algorithms," *J. Comput. and Syst. Sci.*, *13* (1976) 335-379.

6. F.R.K. Chung and A.L. Rosenberg, "Minced trees, with applications to fault-tolerant VLSI processor arrays," *Math. Syst. Th.*, *19* (1986) 1-12.

7. J. Friedman and N. Pippenger, "Expanding graphs contain all small trees," Typescript, IBM Almaden Research Center (1986).

8. D.R. Fulkerson and O.A. Gross, "Incidence matrices and interval graphs," *Pacific J. Math.*, *15* (1965) 835-855.

9. M.R. Garey, R.L. Graham, D.S. Johnson, D.E. Knuth, "Complexity results for bandwidth minimization," *SIAM J. Appl. Math.*, *34* (1978) 477-495.

10. M.R. Garey, D.S. Johnson, L.J. Stockmeyer, "Some simplified *NP*-complete graph problems," *Theoret. Comput. Sci.*, *1* (1976) 237-267.

11. R.J. Lipton and R.E. Tarjan, "A separator theorem for planar graphs," *SIAM J. Appl. Math.*, *36* (1979) 177-189.

12. L. Lovasz, "Coverings and colorings of hypergraphs," pp. 3-12 in *4th Southeast Conf. on Combinatorics, Graph Theory, and Computing*, Utilitas Mathematica Publ., Winnipeg, (1973).

13. C.H. Papadimitriou, "The *NP*-completeness of the bandwidth minimization problem," *Computing*, *16* (1976) 263-270.

14. G.L. Peterson and Y.-H. Ting, "Trade-offs in VLSI for bus communication networks," Tech. Rpt. 111, Univ. Rochester (1982).

15. A.L. Rosenberg, "Routing with permuters: Toward reconfigurable and fault-tolerant networks," Tech. Rpt. CS-1981-13, Duke Univ. (1981).

16. A.L. Rosenberg, "On designing fault-tolerant VLSI processor arrays," pp. 181-204 in *Advances in Computing Research 2*, (F.P. Preparata, ed.) JAI Press, Greenwich, CT, (1984).

17. A.L. Rosenberg, "A hypergraph model for fault-tolerant VLSI processor arrays," *IEEE Trans. Comp.*, *C-34* (1985) 578-584.

18. Y. Shiloach, "A minimum linear arrangement algorithm for undirected trees" *SIAM J. Comput.*, *8* (1979) 15-32.

19. Q. Stout, "Meshes with multiple busses," *27th IEEE Symp. on Foundations of Computer Science* (1986) 264-273.

20. L.G. Valiant, "Universality considerations in VLSI circuits," *IEEE Trans. Comp.*, *C-30* (1981) 135-140.

DEPARTMENT OF COMPUTER AND INFORMATION SCIENCE
UNIVERSITY OF MASSACHUSETTS
AMHERST, MA 01003

Contemporary Mathematics
Volume **89**, 1989

# Competitive Algorithms for On-line Problems [1]

(extended abstract)

*Mark S. Manasse*
DEC Systems Research Center
130 Lytton Avenue
Palo Alto, CA 94301

*Lyle A. McGeoch*
Department of Mathematics
Amherst College
Amherst, MA 01002

*Daniel D. Sleator*
Department of Computer Science
Carnegie Mellon University
Pittsburgh, PA 15213

## 1 Introduction

*On-line problems* are common in computer science and operations research. Any situation in which it is necessary to satisfy some request or answer a query immediately (before future requests are known) requires an on-line solution. Most data structure design problems are on-line, as are many scheduling problems (such as planning the motion of an elevator or the motion of the heads of a disk drive), and many caching problems (such as managing a two-level paged memory system or deciding which blocks to keep in which caches in a multiprocessor caching system).

A new approach to the design and analysis of on-line algorithms was discovered by Sleator and Tarjan [7]. They compared the move-to-front heuristic (MTF) for maintaining a linear search list to other list maintenance heuristics. They found that, for any sequence of requests, the amortized performance of MTF is within a factor of four of the performance of any algorithm, even an *off-line* algorithm that can see all future requests.

Motivated by this work, Karlin, Manasse, Rudolph and Sleator [5] devised caching strategies for snoopy caching systems, and showed that their strategies were also within

[1]This abstract is a summary of our work on competitive algorithms. We have omitted the proofs of most of the theorems. Several of the results of this paper appeared in the second author's PhD thesis [6]. These results include the residue theorem, the residue-based 2-server algorithm, and the balance algorithm for $n-1$-servers. We plan to write a full paper containing all omitted proofs as well as other results.

small constant factors of the optimum off-line strategies for any request sequence. These authors coined the term *c-competitive* to refer to an on-line algorithm whose performance is within a factor of $c$ (plus a constant) of optimum on any sequence of requests. Let $C_A(\sigma)$ and $C_B(\sigma)$ denote the cost of algorithms $A$ and $B$ on request sequence $\sigma$. Algorithm $B$ is then $c$-competitive if it is on-line and if there is a constant $a$ such that:

$$\text{For any algorithm } A \text{ and any sequence } \sigma: \quad C_B(\sigma) \leq c \cdot C_A(\sigma) + a.$$

An algorithm is *competitive* if it is $c$-competitive for some constant $c$. We will say that a competitive algorithm is efficient if the constant $c$ is small.

The study of competitive algorithms is interesting for several reasons. First, efficient competitive algorithms have been discovered for a surprising variety of real problems, and more applications are likely to be discovered. Second, a competitive algorithm can be said to exhibit learning behavior in the following sense: if there is some structure or pattern in the request sequence that makes it possible to process it efficiently, then the competitive algorithm will eventually discover this fact and take advantage of it. Finally, experiments with one competitive algorithm (move-to-front) [2] suggest that the competitiveness of an algorithm is a better indication of how well it will perform in practice than average or worst-case analysis. To even begin to do an average case analysis, some assumption must be made about the distribution of requests. An efficient competitive algorithm obviates the need to make this assumption, because it works well for any distribution.

Given a description of an on-line problem, how can it be determined if there is an efficient competitive algorithm for it? Why is it that for some problems there exist competitive algorithms and for others there do not? To investigate these questions, Borodin, Linial and Saks [1] formalized the definition of an on-line problem and called it a *task system*. Task systems include many of the important on-line problems for which competitive algorithms have been devised. They showed that any symmetric task system has an on-line algorithm competitive within a factor $(2n-1)$ (where $n$ is the size of the task system, not its input). They also exhibited a class of task systems for which there is no competitive algorithm with a smaller ratio than $(2n-1)$. Fortunately, most important on-line problems do not fall into this class of task systems, and the $(2n-1)$ lower bound does not apply to these problems.

In this paper we answer the first question in the above paragraph. We have devised a decision procedure which takes as input the description of a task system, and a constant $c$, and determines if there is a $c$-competitive algorithm for the task system. To obtain this decision procedure we developed the notion of *residues*, and proved the *residue theorem*. This theorem states that if there exists a $c$-competitive algorithm for a task system, then there exists a residue-based $c$-competitive algorithm. Applying this theorem reduces from infinite to finite the number of possible algorithms the decision procedure has to consider. Our decision procedure also constructs the $c$-competitive algorithm if there is one.

We have also used residues to obtain competitive algorithms and lower bounds for a more specialized class of task systems called *server* problems. In a $k$-server problem, the motions of $k$ mobile servers in a graph (with arbitrary distances among the nodes) must be planned on-line under a sequence of requests. Each request requires that some vertex be covered by a server. Planning the motion of a two-headed disk is an example of a 2-server problem. The greedy algorithm for this problem was analyzed in the average case in [3]. Using residues we obtain a new algorithm for the 2-server problem that is 2-competitive.

We have also obtained an $(n-1)$-competitive algorithm for the $(n-1)$-server problem in a graph of $n$ nodes. The factor of $(n-1)$ is the best possible.

Another variant on the server problem is the *server problem with excursions*. Suppose that to process a request, it is not mandatory that a server move to cover it. Instead, the server may send off an assistant (at certain cost) to satisfy the request. We have solved certain special cases of this problem. (Most of the results of Borodin, Linial and Saks follow from our results on the $(n-1)$-server problem.)

Finally, we state two interesting conjectures about competitive algorithms for task systems. These conjectures relate the structure of a task system to the minimum value of $c$ for which a $c$-competitive algorithm exists.

# 2 Task systems in general

A *task system* consists of a set of $n$ states $(1, 2, \ldots, n)$, a set of $m$ tasks $(1, 2, \ldots, m)$, an $n$ by $n$ state transition cost matrix $D = \{d_{ij}\}$, and an $n$ by $m$ task cost matrix $C = \{c_{ik}\}$, $(1 \leq i \leq n, 1 \leq k \leq m)$. The numbers $n$, $m$, and the matrices $D$ and $C$ define the task system.

A sequence of requests $\sigma = \sigma(1), \sigma(2), \ldots \sigma(N)$ is to be processed by the system. Each request is one of the tasks. An algorithm for a task system starts out in state $s(0) = 1$, and chooses which state to use for each of the requests in $\sigma$. Let $s(t)$ be the state used by an algorithm $A$ to process request $\sigma(t)$. The cost incurred by algorithm $A$ in processing $\sigma$ is denoted $C_A(\sigma)$, and is given by:

$$C_A(\sigma) = \sum_{1 \leq t \leq N} d_{s(t-1),s(t)} + \sum_{1 \leq t \leq N} c_{s(t),\sigma(t)}.$$

The first term in this formula is the cost of the state transitions, the second term is the cost of satisfying the requests.

An algorithm for a task system is *on-line with look-ahead one* (or simply on-line) if its choice of what state to use for $\sigma(t)$ depends only on $\sigma(1), \ldots, \sigma(t)$. An algorithm is *on-line with look-ahead zero* if its choice of what state to use for $\sigma(t)$ depends only on $\sigma(1), \ldots, \sigma(t-1)$.

This model is general enough to encompass server problems, server problems with excursions, maintaining a linear search list, maintaining a dynamic search tree, snoopy caching problems, and may others.

## 2.1 Residues

Given a task system and a sequence of requests, dynamic programming can be used to find the sequence of states that minimizes the cost. It works as follows. Let $DP(t, s)$ be the cost of the cheapest method to process the first $t$ requests of $\sigma$ and end in a state $s$. It is easy to compute $DP(0, s)$ for all $s$, and it is easy to compute $DP(t, s)$ for all $s$ given $DP(t-1, s)$ for all $s$. The minimum cost method of processing $\sigma$ is the minimum over all

$s$ of $DP(N, s)$. It is also possible for an on-line algorithm to maintain the $DP$ vector. Of course this does not mean that the algorithm will always be in the state which minimizes $DP(t, s)$. It is more useful for it to maintain a slightly different vector, called the residue vector.

For an algorithm $A$, let $C_A(\sigma(1), \ldots \sigma(t))$ be the cost incurred by $A$ in satisfying the first $t$ requests of $\sigma$. Suppose that $A$ is a $c$-competitive algorithm for a particular task system. Consider the following dynamically changing *residue vector*:

$$RES(t, s) = c \cdot DP(t, s) - C_A(\sigma(1), \ldots \sigma(t)).$$

Because $A$ is $c$-competitive, there is a lower bound on every component of the residue vector. (If there was a sequence of requests that caused a residue component to decrease without bound, then this sequence would disprove the claim that the algorithm is $c$-competitive.)

Just as it is possible for an on-line algorithm to maintain the $DP$ vector, it can maintain the $RES$ vector. It turns out that a $c$-competitive algorithm needs to maintain **only** this vector. Call an on-line algorithm *residue-based* if the only state information it keeps from one request to the next is the residue vector, and the state of the task system. We have the following remarkable theorem.

**Theorem 1** (The Residue Theorem) *If there is a c-competitive algorithm for a task system T, then there exists a residue-based c-competitive algorithm for T.*

*Proof.* Let $A$ be the given $c$-competitive algorithm, and $A'$ be the residue-based one that we seek. Algorithm $A'$ works as follows: A new request $r$ has just been issued, and $A'$ must decide what to do with it. Algorithm $A'$ simulates algorithm $A$ on all sequences of length $0, 1, 2, \ldots$. After each sequence is simulated, $A'$ checks to see if the residues obtained by $A$ in the simulated sequence equal those in the residue memory of $A'$, and if the state of the task system reached by $A$ is the same as that of $A'$. The sequence is said to be *good* if both of these conditions hold. Once $A'$ finds a good sequence, it adds request $r$ to the end of the sequence and determines how $A$ would process $r$ in this new sequence. Algorithm $A'$ processes request $r$ just as $A$ would under these circumstances.

Any residue vector obtained by $A'$ must also be obtained by $A$ under some sequence of requests. The residues of $A$ are bounded below, therefore those of $A'$ are also, and thus $A'$ is $c$-competitive.

To finish the proof we need to show that $A'$ is well defined, that is, it always finds a good sequence. We prove this by induction on the length of the request sequence processed by $A'$. The induction hypothesis is: for any request sequence $\sigma'$ of length $k$, there is some sequence $\sigma$ such that the residue vector obtained by applying $A'$ to $\sigma'$ is exactly the same as that obtained by applying $A$ to $\sigma$, and the state of the task system reached by $A'$ after processing $\sigma'$ is the same as that reached by $A$ after processing $\sigma$. (In other words, $\sigma$ is a good sequence with respect to the residues and state of $A'$ after processing $\sigma'$.)

The induction hypothesis is trivially true for $k = 0$. We shall now prove that the induction hypothesis holds for sequences of length $k+1$ given that it holds for sequences of length $k$. Let $\sigma_1$ be the given sequence of length $k+1$, and let $\sigma'$ be the sequence obtained by deleting the last request $r$ from $\sigma_1$. Let $\sigma$ be the good sequence corresponding to $\sigma'$. By the definition of $A'$, $\sigma$ followed by $r$ must be a good sequence for $\sigma_1$. □

This theorem reduces the space of algorithms to consider when searching for a competitive algorithm. We took advantage of this to find the competitive algorithm for the 2-server problem, which will be described below.

## 2.2 A decision procedure

The residue theorem alone is not enough to decide whether a given task system has a $c$-competitive algorithm for a given constant $c$. We need to limit the search for a $c$-competitive algorithm to a finite number of possibilities. Since the residue vectors are not bounded above, there are an infinite number of them to consider. However, it turns out that if there is a $c$-competitive algorithm, then there is one which only maintains *pseudo-residues*. These vectors are updated just as the residue vector is, except that the algorithm has one additional degree of flexibility: it is allowed to decrease the pseudo-residue values at will. The following theorem shows that the pseudo-residues need not ever get too large or too small.

**Theorem 2** *If there is a c-competitive algorithm for a task system T, then there exists a pseudo-residue-based c-competitive algorithm for T in which the pseudo-residues satisfy the following bounds:*

$$-(2c+1)B \leq PRES(\cdot) \leq (2c+1)B,$$

*where B is an upper bound on $d_{ij}$ for all i and j.*

It is also the case that we can limit the values of pseudo-residues to linear combinations of the $d_{ij}$'s and the $c_{ik}$'s with coefficients of the form $a + b \cdot c$ for integral $a$ and $b$. This observation, combined with the theorem, shows that if all the coefficients and $c$ are rational, then there is a pseudo-residue-based algorithm that can only reach a finite number of different pseudo-residue vectors. It is then easy to actually find the algorithm: exhaustively search all the possible responses an algorithm could make for each possible pseudo-residue vector and state. Each response implies a new state and a new pseudo-residue vector. The theorem shows that if an algorithm is $c$-competitive, there must be a way to choose the responses so that no request sequence leads to a pseudo-residue vector that is too high or too low. This gives a decision procedure for determining if there exists a $c$-competitive algorithm for a given task system with rational coefficients.

Although the decision procedure is exponential in $n$, we believe that it can be implemented far more efficiently than indicated by this analysis, and will actually be useful in obtaining better competitive algorithms for some snoopy caching problems, and for verifying (or disproving) the conjectures at the end of this abstract.

# 3 Server problems

A *k-server problem* is specified by a complete graph $G$ on $n$-nodes, where each edge of $G$ has a length (or distance). The length of the edge $(i,j)$ is denoted $d_{ij}$. These lengths are non-negative, and satisfy the triangle inequality ($d_{ij} \leq d_{ik} + d_{kj}$ for all $i$, $j$, and $k$). The matrix $\{d_{ij}\}$ is called the distance matrix. (The server problem is called symmetric if $\{d_{ij}\}$ is a symmetric matrix.) Let $k$ mobile *servers* occupy the vertices of $G$.

Given a fixed $G$ and $k$, a sequence of requests are to be satisfied. Each request specifies a vertex of $G$ that must be covered by a server. If the requested vertex is already covered by a server, then no action needs to be taken to satisfy the request. If it is not covered, then a server must be moved to the requested vertex. The cost of satisfying the request in this case is the distance moved by the server. The requests must be satisfied in the order in which they occur. An algorithm for the server problem is a method of choosing which server to move after each request.

Because of the flexibility in choosing the number of servers and the distance matrix, server problems are general enough to encompass several important scheduling and caching problems. Here we mention two examples.

**Two-headed disks.** This is the problem of planning the motion of the heads of a two-headed disk along a line, under a sequence of requests. Each request requires that one of the heads be moved to a particular point along the line. The problem is to decide which head to move so that the total head movement is small. This is a server problem in which the distance matrix is that of a set of points arranged in a line. This problem, as well as variants where the servers are in a circle, or on the surface of a sphere were considered by Calderbank, Coffman and Flatto [3,4]. All of these problems are instances of the symmetric 2-server problem.

**Paging problems.** In a two-level memory system there are $k$ pages of fast memory, and a total of $n$ pages of memory. The paging problem is that of deciding which page to take out of fast memory when a new page is needed there. This problem is a thinly-disguised instance of the symmetric $k$-server problem in which all distances are unity. The $n$ nodes of the graph correspond to the $n$ pages of address space, and the $k$ servers occupy the nodes corresponding to pages in fast memory. There are other paging problems in which the costs of moving different pages differ. Such a situation corresponds to a $k$-server problem in which all distances are not equal.

We have obtained several interesting results on competitive algorithms for server problems. First, we prove a lower bound on what the competitive ratio must be.

**Theorem 3** *For any k-server problem, there is no c-competitive algorithm for* $c < k$.

This bound is tight if the following conjecture is true.

**Conjecture 4** *For any* $k \geq 1$, *there is a k-competitive algorithm for any symmetric k-server problem.*

We have proven this conjecture in two special cases, namely $k = 2$ and $k = n-1$. In both cases we explicitly construct the competitive algorithm. We start with the case $k = n-1$.

The *balance algorithm* (*BAL*) for the $k$-server problem is defined as follows. For the server on vertex $i$, let $D_i$ denote the total distance moved by that server since the start of the request sequence. The algorithm will maintain these distances. To process a request at vertex $j$, first determine if $j$ is already covered by a server. If it is, do nothing. If $j$ is not covered, then move the server on vertex $i$ to vertex $j$, where $i$ minimizes the expression $D_i + d_{ij}$. In other words, move the server such that its cumulative cost after the move will be the smallest, among all choices.

The reason we call this the balance algorithm is that it tends to balance the use of all of its servers. Specifically, let 1 be the most recently covered vertex, $n$ be the uncovered vertex, and $i$ be any vertex occupied by a server, then we have:

$$-d_{1i} \leq D_i - D_1 \leq d_{in} - d_{1n}.$$

These inequalities are the first part of a proof of the following theorem:

**Theorem 5** *Algorithm BAL is an* $(n-1)$*-competitive algorithm for the symmetric* $(n-1)$*-server problem on n vertices.*

Unfortunately, the balance algorithm is not competitive when the number of servers is not $n-1$. In attempting to find a 2-competitive algorithm for the 2-server problem, we ruled out *BAL* as well as many other simple approaches. We have obtained a residue-based algorithm *RES* that maintains certain invariants on the residues and chooses which server to move by comparing residues.

**Theorem 6** *Algorithm RES is a 2-competitive algorithm for the symmetric 2-server problem.*

## 3.1 Servers with excursions

Suppose we allow the servers to satisfy a request without actually moving to the requested vertex. We call this type of response an *excursion* because it is natural to think of the server as making an excursion to the requested vertex, satisfying the request, then returning to its starting point. Let $r_{ij}$ be the cost for a server at vertex $i$ to make an excursion to vertex $j$.

A natural example of a server problem with excursions is that of where to locate $k$ firehouses. In this case $d_{ij}$ is the cost of moving the firehouse from $i$ to $j$, and $r_{ij}$ is the cost for the firehouse at $i$ to put out a fire at location $j$. In order to obtain a competitive algorithm for this problem, it is clear that if there are many fires at a particular location, then it will be necessary to move a firehouse there, even if moving a firehouse is very expensive.

The general server problem with excursions seems to be very difficult, but we have obtained a competitive algorithm in one special case.

**Theorem 7** *There is a* $(2n-1)$*-competitive algorithm for the symmetric* $(n-1)$*-server problem with excursions. The factor of* $(2n-1)$ *is the best possible.*

*Proof.* Since all vertices but one are covered by a server, an excursion to a particular vertex always costs the same amount. Let this excursion cost for vertex $i$ be $r_i$.

Given an $(n-1)$-server problem with excursions, we can map this problem into a $(2n-1)$-server problem on a graph with $2n$ vertices. Let the original $n$-vertex graph be $G$, and the new graph of $2n$ vertices be $G'$. We obtain $G'$ from $G$ by making two copies of each vertex of $G$. If $i$ and $i'$ are two vertices of $G'$ that came from the same vertex of $G$, then the distance between them is $r_i$. If $i$ and $i'$ came from different vertices of $G$, then the distance between them in $G'$ is the same as it was in $G$.

There is a one to one correspondence between algorithms solving the $(n-1)$-server problem with excursions in $G$ and the $(2n-1)$-server problem in $G'$. Thus the balance algorithm and the lower bound of the previous section prove the theorem. □

The lower bound in this theorem strengthens that of [1], and the upper bound equals theirs, but is slightly less general since they obtain this bound while allowing the excursion costs to vary over time.

## 3.2 Comparing different numbers of servers

The performance of the best algorithm for the $(n-1)$-server problem could be worse than optimal by a factor of $(n-1)$. This large ratio means that there is a great advantage to being off-line. If we handicap the off-line algorithm by giving it fewer servers, perhaps then the on and off-line algorithms will be more comparable. A similar approach was taken for the paging problem in [7] and for the snoopy caching problem in [5].

We have proven the following theorem relating the performance of the on-line algorithm with $k$ servers to the off-line optimum with $h$ servers, where $h \leq k$.

**Theorem 8** *Let A be an on-line algorithm for the k-server problem on an a graph G, and let B be the optimum off-line algorithm for the h-server problem on the same graph. There exist arbitrarily expensive sequences of requests on which the performance of A is worse than that of B by a factor of at least* $k/(k-h+1)$.

If $k = h$, the bound of this theorem is the same as that of Theorem 3. We conjecture that this bound is tight.

# 4 Conjectures

An interesting phenomenon has emerged from these results on competitive algorithms. The competitive factor, $c$, appears to be relatively independent of the cost transition matrix of the task system, $D$. An example of this phenomenon is the fact that there is a 4-competitive algorithm for maintaining a list in which the cost of swapping two neighboring elements is any positive integer $p$. (The algorithm that achieves this keeps a count in each item of the number of accesses to that item modulo $p$, and moves it to the front when the count reaches 0.) Other examples include many of the results on snoopy-caching, where the algorithms are parameterized by the size of the cache block. The state transition costs are proportional to this block size, but the competitive factor is independent of it. We formalize this conjecture as follows:

**Conjecture 9** (Scaling Conjecture) *Let $T$ be a task system with distance matrix $D$, and let $T'$ be the same task system but with distance matrix $pD$, for some positive integer $p$. If there is a look-ahead zero c-competitive algorithm for $T$ then there is one for $T'$.*

An even stronger conjecture appears to be true. This one is motivated by the results concerning server problems, and is more speculative.

**Conjecture 10** (Independence Conjecture) *Let $T$ and $T'$ be two task systems with distance matrices $\{d_{ij}\}$ and $\{d'_{ij}\}$ respectively. Suppose that $d'_{ij} \geq d_{ij}$ for all $i$ and $j$. If there is a look-ahead zero c-competitive algorithm for $T$ then there is one for $T'$.*

# References

[1] Borodin, A., Linial, N., and Saks, M. An optimal online algorithm for metrical task systems. In *Proceedings of the 19th ACM Symposium on Theory of Computing*, pages 373–382, New York, 1987.

[2] Bentley, J. L. and McGeoch, C. C. Amortized analyses of self-organizing sequential search heuristics. *Communications of the ACM*, 28(4):404–411, April 1985.

[3] Calderbank, A. R., Coffman, Jr., E. G., and Flatto, L. Sequencing problems in two-server systems. *Mathematics of Operations Research*, 10(4):585–598, November 1985.

[4] Calderbank, A. R., Coffman, Jr., E. G., and Flatto, L. Sequencing two servers on a sphere. *Commun. Statist.-Stochastic Models*, 1(1):17–28, November 1985.

[5] Karlin, A. R., Manasse, M. S., Rudolph, L., and Sleator, D. D. *Competitive Snoopy Caching*. Computer Science Technical Report CMU-CS-86-164, Carnegie Mellon University, 1986.

[6] McGeoch, L. A., *Algorithms for Two Graph Problems* PhD Thesis, Department of Computer Science, Carnegie Mellon University, 1987.

[7] Sleator, D. D. and Tarjan, R. E. Amortized efficiency of list update and paging rules. *Communications of the ACM*, 28(2):202–208, February 1985.

Contemporary Mathematics
Volume 89, 1989

# ON RECOGNIZABILITY OF PLANAR GRAPHS

Larry I. Basenspiler

ABSTRACT. The object of this paper is to show that planarity of a graph can be derived from a deck of its non-isomorphic elementary contractions. The same is deduced for any graph and a deck of its non-isomorphic edge-deleted subgraphs which is an extension of a similar result, with some constraints, previously obtained by Fiorini.

1. The graph theoretic notation is essentially that of Harary [5]. Some basic definitions and results are briefly reviewed below, and others are introduced as necessary. Let $G(n,m)$ denote a finite simple graph with vertex set $V(G)$, edge set $E(G)$, $n$ vertices $v_1, v_2, \dots, v_n$ and $m$ edges $e_1, e_2, \dots, e_m$. A subgraph of G obtained by deleting a vertex v (edge e) will be referred to as a vertex-deleted (edge-deleted) subgraph and denoted by $G-v$ $(G-e)$. A graph obtained from G by identifying a pair of adjacent vertices connected by edge e, with subsequent removal of a loop and possible multiple edges, is called an elementary contraction $G\backslash e$ of G. A graph obtained from G by any combination of elementary contractions is called a contraction of G, and a graph obtained by any combination of edge-deletions and elementary contractions is called a minor of G. A reconstruction (edge-reconstruction) of a graph G is a graph H such that $V(H) = V(G)$ $(E(H) = E(G))$ and $H-v = G-v$ $(H-e = G-e)$ for all $v \in V(G)$ ( for all $e \in E(G)$ ). A C-reconstruction of a graph G is a graph H such that $E(H) = E(G)$ and $H\backslash e = G\backslash e$ for all $e \in E(G)$. G is reconstructible if every reconstruction of G is isomorphic to G. The Reconstruction Conjecture states that a graph is reconstructible, uniquely up to isomorphism, from its proper subgraphs [6] ( its elementary

1980 Mathematics Subject Classification (1985 Revision) 05C60.

contractions [1,3]).

Along with the attempts to prove the Reconstruction Conjecture directly, there exists another approach to the problem, namely: the reconstruction of parameters, i.e. investigation into a question as to what functions of a graph G take the same value on all reconstructions ( not necessarily isomorphic to each other ) of G. This implies a problem which will be referred to as "recognizability" ( with the prefixes V-, E-, and C- for vertex-, edge-subgraphs, and elementary contractions, respectively): A class of graphs is recognizable if, for each graph G in $\mathcal{G}$, every reconstruction of G is also in $\mathcal{G}$. In this paper we prove that planar graphs are C- and E-recognizable. The following criterion of planarity of a graph will be used herein ( Kuratowski-Wagner [5]): a graph is planar if and only if it does not have a subgraph contractible to $K_5$ ( Fig. 1a ) or $K_{3,3}$ ( Fig. 1b ), or in other words, it does not contain forbidden minors $K_5$ or $K_{3,3}$.

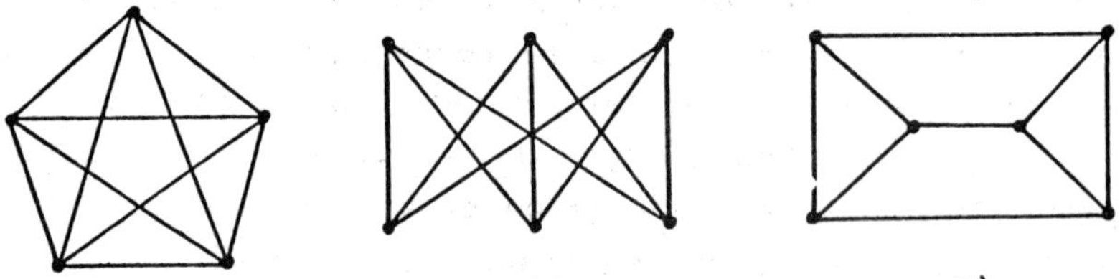

a) b) c)

Fig. 1

2. In this section we show that planar graphs are C-recognizable. As it has been shown [2], Kuratowski's criterion of planarity of a graph can be re-formulated in terms of the contractions of the graph only. That is, consideration of the subgraphs of a graph is unnecessary for planarity testing.

LEMMA 1 ( Main Theorem in [2]). A graph is planar if and only if it is not contractible to $K_5$ or to a graph $G(6,m)$ ( $m > 8$ ) satisfying one of the following conditions:

1. $m = 9$ and all elementary contractions of G have 8 edges;
2. $m = 10$ and no elementary contraction of G has 7 edges;
3. $m = 11$ and no 3 elementary contractions of G have 9 edges;
4. $m = 12$, at least one elementary contraction of G has 7 edges, and no contraction has 8 edges.

THEOREM 1. A graph of order at least 7 is planar if and only if every one of its elementary contractions is planar.

Proof. By lemma 1, one can assert that a graph on $n > 6$ vertices is planar if and only if it is not contractible to certain graphs on 6 vertices. Let G be a non-planar graph. Then at least one of its elementary contractions does not satisfy the conditions of lemma 1. Since $n > 6$, it must be reflected in at least one of its elementary contractions, which proves the result.

Moreover, in proving theorem 1, it obviously suffices to consider only nonisomorphic elementary contractions of a graph.

COROLLARY 1. A planar graph is C-recognizable from the set of its nonisomorphic elementary contractions.

Proof. The conditions of lemma 1 provide an easy planarity test for graphs of order 6 ( 6-graphs). Thus, theorem 1 and lemma 1 combined produce the desired result.

2. In this section we show that planar graphs are E-recognizable. Hereafter, an edge-deleted subgraph will be called simply a subgraph. Following Manvel [7], denote the set of nonisomorphic subgraphs $G - e_i$ of a graph G by Q. If such a property as planarity can be inferred from Q, it is said that planarity of G is Q-recognizable. Fiorini [4] has proved that planar connected graphs of order at least 7 and minimum valency at least 3 are recognizable.

LEMMA 2 (Theorem 2 in [4] ). A graph of order at least 7 and minimum valency at least 3 is planar if and only if every edge-deleted subgraph is planar.

It is easy to see that Fiorini's result can be extended to the case when only the set Q is known.

THEOREM 2. A planar graph is Q-recognizable.

Proof. If a graph has at least one nonplanar subgraph ( and thereby is itself nonplanar ), we shall refer to it, in light of lemma 2, as an F-graph. Obviously if a nonplanar graph is an F-graph, its nonplanarity is Q-recognizable due to Kuratowski's criterion. The fact that before contracting a graph one can remove at least one edge underlies the proof of lemma 2. The only nonplanar graphs, which are not the subject of the restriction of lemma 2, are 5- and 6-graphs, and those which themselves are contractible to $K_{3,3}$ or $K_5$. We prove here that they are also recognizable. Manvel [7] has shown that the degree sequence S(G) of G can be derived from the set Q of G. Let H be a graph obtained from $G-e \in Q(G)$ by adding an edge so that $S(H) = S(G)$, then H is said to be "an easy reconstruction by degree sequence" or, for brevity, "an ED-reconstruction" of a graph G. If H is isomorphic to G then the ED-reconstruction is said to be a true reconstruction. Consider all nonplanar 6-graphs which have a subgraph isomorphic to $K_{3,3}$. All of these graphs ( which can be found in Appendix 1 in [5] ), except for two in Fig. 1b and 1c, are F-graphs. The two exceptional graphs, however, can be readily ED-reconstructed uniquely up to isomorphism as regular graphs.

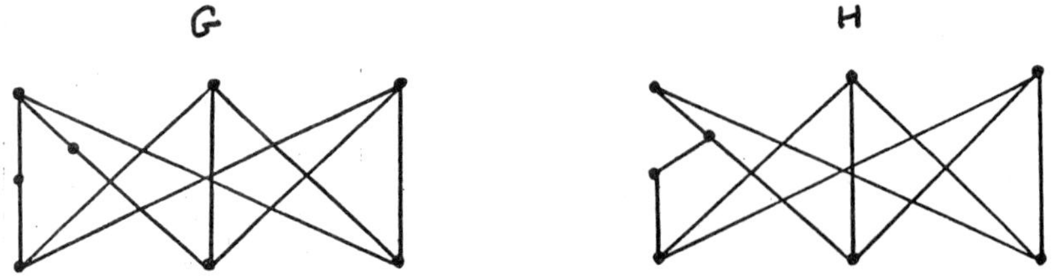


Fig. 2

Notice that 6-graphs contractible to $K_5$ are either F-graphs or their ED-reconstruction is a true reconstruction. That is, planar 6-graphs are Q-recognizable. Let G be a nonplanar m-graph ($m > 6$ ) homeomorphic to $K_{3,3}$. Then G has exactly 6 vertices of degree 3 ( 3-vertices) and all other vertices are 2-vertices. If at least two 3-vertices are connected by a path of length > 2 , then one can find in the set Q(G) a subgraph with two 1-vertices, connect those vertices with an edge, and that ED-reconstruction will induce the true reconstruction. This leaves for analysis only the graphs with at most 15 vertices induced by all possible insertions of a vertex of degree 2 into edges of $K_{3,3}$. Call such a subdiveded edge a 2-path. Notice that when a graph has only one 2-path, the obvious ED-reconstruction again induces the true reconstruction. When two 2-paths belong to only one 3-vertex, as in Fig. 2a, an ED-reconstruction from a subgraph with a 1-vertex

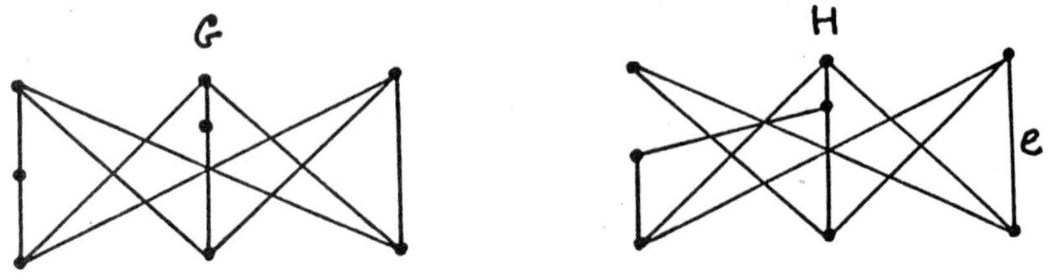


Fig. 3

may produce graphs G and H in Fig. 2 which are isomorphic, and this ED-reconstruction is the true reconstruction. In the case

when two 2-paths belong to the different 3-vertices, an ED-reconstruction may produce planar and nonplanar versions of a graph ( graphs G and H in Fig. 3 ). However, H - e has three 3-vertices lying on a path of length 2 whereas no subgraph in Q(G) possesses this property. That is, Q(G) = Q(H), and appropriately chosen subgraph of G ( or H ) can produce a true reconstruction. Direct analysis of the cases of more than two 2-vertices (that is, 3,...,9), as well as the case of the graphs contractible to $K_5$, is simple, though tedious, and is omitted here. It shows that nonplanar graphs which are not F-graphs are not only recognizable but moreover reconstructible. Q.E.D.

The author is grateful to Margie Brooks for numerous improvements in style and presentation.

## BIBLIOGRAPHY

1. Basenspiler L. I. and Choizonova K. L., "On Reconstruction Problem for Graphs," Proceedings in Applied Mathematics and Cybernetics, Siberia Section of Academia of Science of USSR, State Research Institute of Technical and Scientific Information, 5285-72 DEP (1973), 49-55.
2. Basenspiler L. I., "A Note on Forbidden Minors." The 250th Anniversary Conference on Graph Theory. Indiana University-Purdue University, Fort Wayne, Indiana, March 1986.
3. Bhave V. N., Kundu S. and Sampathkumar E., "Reconstruction of a tree from its homomorhic images and other related transforms. J. Combin. Theory Ser. B. 20, # 2, (1976), 117-123.
4. Fiorini S., "On the Edge-Reconstruction of planar graphs." Math. Proc. Camb. Phil. Soc. 83, (1978), 31-35.
5. Harary F., Graph Theory, Addison-Wesley,Reading, MA, 1969.
6. Harary F., "On the Reconstruction of a Graph from a collection of subgraphs", Theory of Graphs and Applications (M. Fiedler, ed.) Academic Press, New York, (1964), 47 - 52.
7. Manvel B., "On Reconstructing Graphs from Their Sets of Subgraphs." Journal of Combin. Theory (B) 21, (1976), 156-165.

DEPARTMENT OF COMPUTER SCIENCE
NORTHERN ILLINOIS UNIVERSITY
DE KALB, ILLINOIS 60115

Current address:
Division of Computer & Information Sciences,
University of South Alabama,
Mobile, AL 35588

Contemporary Mathematics
Volume **89**, 1989

# COMBINATORIAL COMPUTATION OF MODULI DIMENSION OF NIELSEN CLASSES OF COVERS

*Emphasis on the solvable cover case with historical comments from Zariski Vol. 3*

*Mike Fried, UC Irvine and U. of Florida*

**Abstract:** Consider a rational function field $\mathcal{C}(x)$ in one variable. There have been quite a number of attempts to use Riemann's existence theorem to organize both the lattice of subfields and the algebraic extensions of it. This exposition describes a further attempt that includes exposition on ground (§2) covered sporadically by Zariski [Z]. A rough phrasing of the particular problem: For each nonnegative integer $g$ describe explicitly all of the ways that the function field of the "generic curve" of genus $g$ contains $\mathcal{C}(x)$ (§1).

Although a general program has been envisioned by John Thompson (c.f. §2.2), we narrow to the case where $g \geq 2$ and the containment of fields gives a solvable Galois closure. This alone illustrates that Zariski's most definitive conjecture touching on this is wrong (§2.3). Theorem 3.5 gives a presentation of the fundamental group $\pi_1(X)$ and 1st homology $H_1(X, \mathbf{Z})$ of a Riemann surface $X$ appearing in a (not necessarily Galois) cover $X \to \mathbf{P}^1_x$ of the sphere in terms of branch cycles for the cover. In particular this offers an action of the Hurwitz monodromy group $H(r)$ on $H_1(X, \mathbf{Z})$ where $r$ is the number of branch points of the cover. The remainder of §3 interprets the dimension of the image of the deformations of the cover in the moduli space of curves of genus $g = g(X)$ in terms of this group action.

## . STATEMENT OF THE PROBLEM AND OUTLINE OF THE SECTIONS

### 1. FAMILIES OF POLYNOMIALS WITH GIVEN MONODROMY GROUP

e way to give an (irreducible) algebraic curve is to give a polynomial (irreducible) in two iables $f(x,y) \in \mathcal{C}[x,y]$ where $\mathcal{C}$ denotes the complex numbers. Then the curve in question is

$$\{(x,y) \mid f(x,y) = 0\} \stackrel{\text{def}}{=} X.$$

is curve may, however, have singular points: points $(x_0, y_0) \in X$ for which $\frac{\partial f}{\partial x}$ and $\frac{\partial f}{\partial y}$ evaluated $(x_0, y_0)$ are both 0. Furthermore, we are missing the points at infinity obtained by taking the sure of $X$ in the natural copy of projective 2-space $\mathbf{P}^2$ that contains the affine space $\mathbf{A}^2$ with iables $x$ and $y$. (And these points, too, might be singular.)

---

AMS Subject classification: 12F05, 14H40, 14D20, 14E20, 20F36

Keywords: Extensions of $\mathcal{C}(z)$; algebraic moduli problems; Riemann's existence theorem; obian varieties; Artin braid group; Hurwitz monodromy group; Zariski's collected works.

Support gratefully accepted from National Science Foundation grant DMS–8702150

After this opening section we will assume that our algebraic curves $X$ don't have these defects; they will be projective nonsingular curves, so we may not be able to regard them as given by a single polynomial in 2-space. But the essential ingredient of this presentation, represented by the $x$-coordinate will still be there.

That is, we have a *covering* map

$$\{(x,y) \mid f(x,y)=0\} \to \mathbf{P}^1_x \stackrel{\text{def}}{=} \mathcal{C} \cup \infty \quad \text{or} \quad X \to \mathbf{P}^1_x \tag{1.1}$$

given by projection of the point $(x,y)$ onto its first coordinate. The *monodromy group* of this cover is defined to be the Galois group $G$ of the Galois closure of the field extension $\mathcal{C}(X)/\mathcal{C}(x)$ where $\mathcal{C}(X)$ denotes the quotient field of the ring $\mathcal{C}[x,y]/(f(x,y))$. In the sequal we will denote this Galois closure by $\widehat{\mathcal{C}(X)}$ or by the geometric version $\hat{X}$, the smallest Galois cover of $\mathbf{P}^1_x$ that factors through $X \to \mathbf{P}^1_x$. Note that in this situation $G$ automatically comes equipped with a transitive permutation representation $T : G \to S_n$. Denote the stabilizer in $G$ of an integer (say, 1) by $G(T)$. Also, for later reference we point out that $T$ is *primitive* (i.e., there are no proper groups between $G$ and $G(T)$) if and only if there are no proper fields between $\mathcal{C}(X)$ and $\mathcal{C}(x)$ (equivalently, no proper covers fitting between $X \to \mathbf{P}^1_x$).

Actually, the problem of concern doesn't deal with one polynomial at a time, but rather with a parametrized family of them. We give the technical details for this in §1.2, but for a statement of the main problem it suffices to think of the coefficients of $f(x,y)$ lying in a field $F$, finitely generated over the rationals $\mathbf{Q}$. The problem comes when we simultaneously want to declare further properties of the (ramified) cover $X \to \mathbf{P}^1_x$ and for the field $F$. Here is the naive version of the constraints that we impose in terms of a priori given data, a group $G$ and a nonnegative integer $g$:

(1.2) a) The monodromy group of the cover $X \to \mathbf{P}^1_x$ is equal to the group $G$; and

b) As we run over all specializations of the field $F$ in the complex numbers, the field $\mathcal{C}(X)$ runs over "almost all" fields of functions of Riemann surfaces of genus $g$.

The phrase "almost all" means for all but a codimension 1 algebraic subset of the moduli space $\mathcal{M}_g$ of Riemann surfaces of genus $g$. The existence of such an (irreducible) algebraic variety and its properties ([M; Lecture II]—c.f. §2) is, of course, no triviality. Indeed, it is the abstractness of this object that causes all of our problems when we want to find out for which pairs $(g,G)$ there exists such a polynomial $f(x,y)$ with coefficients in such a field $F$. The function fields of these polynomials depend only on the isomorphism class of the representing Riemann surface. Denote the points $\boldsymbol{m} \in \mathcal{M}_g$ whose representing Riemann surfaces have the same function field as such a polynomial $f$ (as in (1.1)) by $\mathcal{M}_g(G)$. We may rephrase the conditions of (1.2) in the following form:

**Question 1.1:** *For which $(g,G)$ is $\mathcal{M}_g(G)$ a Zariski open subset of $\mathcal{M}_g$?*

For fixed $g$ denote the collection of groups for which the conclusion of the question is affirmative by $\mathcal{G}_g$ and denote the subset of solvable groups by $\mathcal{G}_g(\text{sol})$. The Main Theorem of [FrG] considers $\mathcal{G}_g(\text{sol})$ in the case that $g \geq 2$. Indeed, for a fixed $g$ in order to show that $\mathcal{G}_g(\text{sol})$ is empty it suffices to show that the subset $\mathcal{G}_g(\text{prim}) \cap \mathcal{G}_g(\text{sol})$ consisting of primitive groups is empty. The Main Theorem of [FrG] gives the following.

**Theorem 1.2:** *For $g \geq 7$ $\mathcal{G}_g(sol)$ is empty; for $3 \leq g \leq 6$, $\mathcal{G}_g(prim) \cap \mathcal{G}_g(sol)$ consists of just $S_3$ and $S_4$; and for $g = 2$, in addition to $S_3$ and $S_4$, $\mathcal{G}_g(prim) \cap \mathcal{G}_g(sol)$ is a subset of this list:*

$$G = D_{10}; \qquad (\mathbf{Z}/3)^2 \times^s D_8; \qquad (\mathbf{Z}/3)^2 \times^s GL(2,3); \quad \textit{and} \tag{1.3}$$

$$(\mathbf{Z}/2)^2 \times (\mathbf{Z}/2)^2 \times^s ((S_3 \times S_3) \times^s (\mathbf{Z}/2)).$$

The most exciting *mathematical* considerations of [FrG] revolve around deciding which of the members of (1.3) are actually in $\mathcal{G}_g(\text{prim})$. This is a special case of the computation of e *moduli dimension* of a *Nielsen class* (§1.2). The *Hurwitz monodromy group*, a quotient of e Artin braid group plays the key role in reducing the problem to a computation in pure group eory. The computations, however, are difficult even if of general interest. Some of the list (1.3) as been eliminated by them, but there are two groups that are still in question at the time of is writing. Furthermore, the ideas can be applied to many problems, so it would be a shame if ke minded calculations turned out to be unfeasible (§3).

In particular, Thompson's program (§2.2) conjectures that for fixed $g$, excluding $A_n$, $= 5, 6, \ldots$, there are only finitely many simple groups that appear as composition factors of onodromy groups of covers by a Riemann surface of genus $g$. Therefore we illustrate further oup theoretical computational difficulties on the problem of deciding for fixed $g$ those $n$ for hich $(g, A_n)$ gives an affirmative answer to Question 1.1.

The most interesting *historical* considerations of [FrG] are best summarized by noting at Zariski considered almost all aspects of the problem—unbeknownst to the authors of [FrG] the appearance of the first draft of their paper—among a considerable subset of the papers in e 3rd volume of his collected works [Z]. In fact he knew everything in Theorem 1.2. except list .3). But, in the course of his formulation of a special case of the "moduli dimension problem" conjectured results from which one would conclude that all of list (1.3) is in $\mathcal{G}_g(\text{prim}) \cap \mathcal{G}_g(\text{sol})$, ntrary to our statement above. Also, he never explored the different ways that $S_3$ and $S_4$ belong this list. More precisely, in the phraseology of §1.2: For which *Nielsen classes* $\mathbf{C}$ is $\text{Ni}(\mathbf{C})_T^{ab}$ of ll *moduli dimension* where $T$ is the standard representation of either $S_3$ or $S_4$. The minimal teger $n$ for which $(g, S_n)$ gives an affirmative answer to Question 1.1 is $n = [\frac{g+3}{2}]$ where [ ] notes the "greatest integer" function [KL] (c.f. §2.1). The §1.2 formulations show that this is e obvious first case of the problem of computation of moduli dimension of a Nielsen class: the se when the group is $S_n$ and the conjugacy classes $\mathbf{C} = (\mathrm{C}_1, \ldots, \mathrm{C}_r)$ in $G = S_n$ are each the njugacy class of a 2-cycle. In §2 we explore the relation between [Z] and [FrG], with pointed marks about [AM].

A paraphrase of Theorem 1.2 might start like this.

**tatement 1.3:** *The generic curve of genus $g > 6$ is not uniformized by radicals.*

deed, this was the first draft title of [FrG], which turned out to be essentially the English version the Italian title of item [8] of [Z] (c.f. §2.1). I think that those who are comfortable with the assical treatment of algebraic geometry will have no difficulty with the limits of this result. ut it is illuminating to point out that at this time it is not known for any $g$ whether or not solvable ${}_G\mathcal{M}_g(G)$ is dense in $\mathcal{M}_g$. In particular, it is (vaguely) possible for some $g > 6$ that ch curve of genus $g$ defined over $\bar{\mathbf{Q}}$, the algebraic closure of the rationals, has *some* map to $\mathbf{P}_x^1$ hose monodromy group is solvable.

**cknowledgements:** It was John Ries who realized that a number of points of [FrG] are lated to [Z], and who also, in looking back, had the first counterexamples to the conjecture of m [18] (see §2) of [Z]. A number of readers of a draft of this article have warned me that while is historically conservative to be cavalier about the definition of "generic," the modern reader ll not allow such liberties. Here, at least, I have tried to keep the reader's comfort in mind. seems, however, inevitable that some intended readers might be unwilling to suspend concern at they haven't the background to visualize a hard core algebraic geometry object like $\mathcal{M}_g$ or (C)$_T$. For those willing to travel adventurously in the direction of a proferred arrow, I give a ide to their properties. Other than that I can only say that all pedantry is unintended.

## .2. NIELSEN CLASSES AND THE HURWITZ MONODROMY GROUP

Suppose that we are given a finite set $\{x_1, \ldots, x_r\}$ of distinct points of $\mathbf{P}_x^1$. For any ment $\boldsymbol{\sigma} \in S_n^r$ denote the group generated by its coordinate entries by $G(\boldsymbol{\sigma})$. We recall the classical classification data for the connected (ramified) degree $n$ covers of the $x$-sphere. Consider $\phi : X \to \mathbf{P}^1_x$, ramified only over $\boldsymbol{x}$ up to the relation that regards $\phi : X \to \mathbf{P}^1_x$ and $\phi' : X' \to \mathbf{P}^1_x$ as equivalent if there exists a homeomorphism $\lambda : X \to X'$ such that $\phi' \circ \lambda = \phi$. These equivalence classes are in one-one correspondence with

$$(1.4) \qquad \{\boldsymbol{\sigma} = (\sigma_1, \dots, \sigma_r) \in S_n^r \mid \sigma_1 \cdots \sigma_r = 1,\ G(\boldsymbol{\sigma}) \text{ is a transitive subgroup of } S_n\}$$

modulo the relation that regards $\boldsymbol{\sigma}$ and $\boldsymbol{\sigma}'$ as equivalent if there exists $\gamma \in S_n$ with $\gamma\boldsymbol{\sigma}\gamma^{-1} = \boldsymbol{\sigma}'$. This correspondence goes under the heading of *Riemann's existence theorem*. The collection of ramified points $\boldsymbol{x}$ will be called the branch points of the cover $\phi : X \to \mathbf{P}^1_x$. (In most practical situations we shall mean that there truly is ramification over *each* of the points $x_i$, $i = 1, \dots, r$.)

Our next step is to generalize Riemann's existence theorem to a combinatorial group situation that allows us to consider the covers above, not one at a time, but as topologized collections of families: the branch points $\boldsymbol{x}$ run over the set $(\mathbf{P}^1_x)^r \setminus \Delta^r$ with $\Delta_r$ the $r$-tuples with two or more coordinates equal. The key definition is of a *Nielsen class*.

Suppose that $T : G \to S_n$ is any faithful transitive permutation representation of a group $G$. Let $\mathbf{C} = (\mathrm{C}_1, \dots, \mathrm{C}_r)$ be an $r$-tuple of conjugacy classes from $G$. It is understood in our next definition that we have fixed the group $G$ before introducing conjugacy classes from it.

**Definition 1.4:** The Nielsen class of $\mathbf{C}$ is

$$\mathrm{Ni}(\mathbf{C}) \overset{\mathrm{def}}{=} \{\boldsymbol{\tau} \in G^r \mid G(\boldsymbol{\tau}) = G \text{ and there exists } \beta \in S_r \quad \text{such that } \tau_{\beta(i)} \in \mathrm{C}_i,\ i = 1, \dots, r\}.$$

We always assume that any given Nielsen class under consideration is nonempty—but, of course, this must be checked in each case. Also, for simplicity we assume that $\mathrm{C}_i$ is not the conjugacy class of the identity, $i = 1, \dots, r$.

Relative to *canonical* generators $\bar\sigma_1, \dots, \bar\sigma_r$ (see Figure 1) of the fundamental group $\pi_1(\mathbf{P}^1_x - \boldsymbol{x}, x_0)$, we say that a cover ramified only over $\boldsymbol{x}$ is in $\mathrm{Ni}(\mathbf{C})$ if the classical representation of the fundamental group sends the respective canonical generators to an $r$-tuple $\boldsymbol{\sigma} \in \mathrm{Ni}(\mathbf{C})$. What we would like to have is a total family of covers of $\mathbf{P}^1_x$ representing these equivalence classes. There are subtleties to forming this—even talking about it. Our next simplifying assumption on $G$ holds for all examples of this paper.

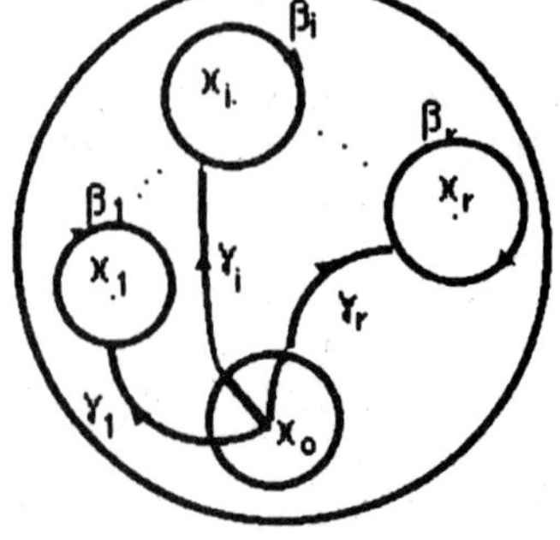


**Figure 1:**
**Sample Bouquet**
$\gamma_i\beta_i\gamma_i^{-1}$ represents one of the oriented generators $\bar\sigma_i$ of the $r$-punctured sphere.

Assume that the permutation representation $T : G \to S_n$ has the property that

$$(1.5) \qquad \text{the centralizer } \quad \mathrm{Cen}_{S_n}(G) \quad \text{of } G \text{ in } S_n \text{ is trivial.}$$

For example, any primitive subgroup of $S_n$ satisfies (1.5). As all of §4 of [Fr,1] makes clear, the practical use of families without condition (1.5) is difficult, but not impossible.

Each permutation representation $T : G \to S_n$ provides us with an important equivalence relation on $\mathrm{Ni}(\mathbf{C})$. Consider the normalizer $N_{S_n}(G)$ (or $N_T(G)$) of $G$ in $S_n$. The subgroup of the normalizer that consists of elements that permute the conjugacy classes $\mathrm{C}_i$, $i = 1, \dots, r$ (under conjugation) is denoted $N_T(\mathbf{C})$. The quotient of $\mathrm{Ni}(\mathbf{C})$ by this group called the *absolute Nielsen classes* (relative to $T$) and it is denoted by $\mathrm{Ni}(\mathbf{C})^{ab}_T$.

We now define the Hurwitz monodromy group $H(r)$—a quotient of the Artin braid group c.f. §2.4 for Zariski's research into this group). The generators $Q_1, \dots, Q_{r-1}$ of $H(r)$ satisfy the ollowing relations:

1.6) a) $Q_i Q_{i+1} Q_i = Q_{i+1} Q_i Q_{i+1}$, $i = 1, \dots, r-2$;
b) $Q_i Q_j = Q_j Q_i$ for $1 \le i < j - 1 \le r - 1$; and
c) $Q_1 Q_2 \cdots Q_{r-1} Q_{r-1} \cdots Q_1 = 1$.

Relations (1.6) a) and b) alone give the braid group. Their "strings" are not directly part of our setup. It is relation (1.6) c) that truly indicates our involvement with projective lgebraic geometry; the Artin braid group is the fundamental group of $\mathbf{A}^r - D_r$ while the Hurwitz nonodromy group is the fundamental group of $\mathbf{P}^r - D_r$. Here $D_r$ is the classical discriminant ocus in the respective spaces. The natural embedding of $\mathbf{A}^r$ in $\mathbf{P}^r$ gives the natural surjective omomorphism from the braid group to the monodromy group.

From the relations we compute that $H(r)$ acts on the absolute Nielsen classes by extension f the following formula:

1.7) $$(\tau_1, \dots, \tau_r)Q_i = (\tau_1, \dots, \tau_{i-1}, \tau_i \tau_{i+1} \tau_i^{-1}, \tau_i, \tau_{i+2}, \dots, \tau_r).$$

n the notation above we say that $\phi_T : X_T \to \mathbf{P}^1_x$ is in the absolute Nielsen class $\mathrm{Ni}(\mathbf{C})^{ab}_T$. In nany contexts it would be impossible to drop the subscript $T$ without confusion. But such is nlikely to occur in this paper. Therefore we drop the subscript $T$ quite often.

Each absolute Nielsen class $\mathrm{Ni}(\mathbf{C})^{ab}_T$ defines a *moduli space* $\mathcal{H}(\mathbf{C})_T$ of covers $\phi_T : X_T \to$ $\mathbf{P}^1_x$ of degree equal to $\deg(T)$ in that Nielsen class [Fr,1; §4]. Each point of $\mathcal{H}(\mathbf{C})_T$ corresponds o exactly one equivalence class of covers of $\mathrm{Ni}(\mathbf{C})^{ab}_T$. To explain the meaning of this we need a ttle notation to explain *families* of covers in $\mathrm{Ni}(\mathbf{C})^{ab}_T$.

Indeed, such a family $\mathcal{F}$ consists of a *parameter space* $\mathcal{H}$, a *total space* $\mathcal{T}$ and a map $: \mathcal{T} \to \mathcal{H} \times \mathbf{P}^1_x$, all complex manifolds, with this property: For each $\boldsymbol{m} \in \mathcal{H}$ the restriction of $\Phi$ o the fiber

$$\mathcal{T}_{\boldsymbol{m}} \stackrel{\text{def}}{=} \{t \in \mathcal{T} \mid \Phi(t) \in \boldsymbol{m} \times \mathbf{P}^1_x\}$$

resents it as a cover in the absolute Nielsen class $\mathrm{Ni}(\mathbf{C})^{ab}_T$. For brevity denote this cover by $\mathcal{F}_{\boldsymbol{m}}$, nd its equivalence class (representing a point of $\mathcal{M}_g$) by $[\mathcal{F}_{\boldsymbol{m}}]$. The main moduli space property that the natural map

.8) $$\Psi(\mathcal{H}, \mathcal{M}_g) : \mathcal{H} \to \mathcal{M}_g \quad \text{by} \quad \boldsymbol{m} \to [\mathcal{F}_{\boldsymbol{m}}]$$

complex analytic (actually algebraic as both spaces are quasi-projective varieties).

A final point: If (1.5) holds, then there is a unique family (up to the obvious equivalence) $(\mathbf{C})_T$,

.9) $$\Phi(\mathbf{C})_T : \mathcal{T}(\mathbf{C})_T \to \mathcal{H}(\mathbf{C})_T \times \mathbf{P}^1_x ,$$

ıch that for each $\boldsymbol{m} \in \mathcal{H}(\mathbf{C})_T$ the restriction of $\Phi(\mathbf{C})_T$ to the fiber $\mathcal{T}(\mathbf{C})_{T,\boldsymbol{m}}$ presents it as a over in the absolute Nielsen class $\mathrm{Ni}(\mathbf{C})^{ab}_T$ (e.g., [Fr,1; §4]). The parameter variety $\mathcal{H}(\mathbf{C})_T$ ties us ack to the elementary phrasing of §1.1. We are forced to consider the function field $\mathcal{C}(\mathcal{H}(\mathbf{C})_T)$ $\mathcal{H}(\mathbf{C})_T$ for suitable $\mathbf{C}$ as a candidate for the field $F$ in (1.2). And if for no suitable Nielsen ass does this choice of $F$ work, then Question 1.1 has a negative answer for the given $(g, G)$.

**efinition 1.5:** The *moduli dimension* of the Nielsen class $\mathrm{Ni}(\mathbf{C})^{ab}_T$ is the dimension of the nage of the morphism $\Psi(\mathcal{H}(\mathbf{C})_T, \mathcal{M}_g)$. We say that $\mathrm{Ni}(\mathbf{C})^{ab}_T$ is of *full* moduli dimension if this ap is dominant (i.e., generically surjective). In other words, if the range has dimension $3g - 3$ esp., $g$) if $g \ge 2$ (resp., $g$=0 or 1).

The case when $G = S_n$ and $\mathbf{C}$ consists of just the conjugacy class of 2-cycles should be garded as the classical case of this problem. Covers in this Nielsen class are said to be *simple ranched covers*. The result in this case is that the $n$ for which $\mathrm{Ni}(\mathbf{C})^{ab}_T$ has full moduli dimension are exactly the $n \geq [\frac{g+3}{2}]$. This is hardly trivial (c.f. §1.1). Indeed, while Zariski clearly "knew this during the writing of his papers, [AM] regards it as still open until [KL] (§2.1). Neverthelesthere is a principle— known to the ancients—that applies to this situation.

**Principle 1.6:** *If any Riemann surface $X$ of genus $g$ has a covering $X \to \mathbf{P}^1_x$ of degree $n$, the some Riemann surface $X'$ of genus $g$ appears as a simple branched cover of $\mathbf{P}^1_x$ of degree $n$.*

Since nothing like this holds for Nielsen classes in general (c.f. Statement 2.16). we fe that some additions to the classical geometry ideas of, say [KL] and [ArC], would be required decide the moduli dimension of $\mathrm{Ni}(\mathbf{C})^{ab}_T$ in general. The goal of [FrG] is to return this proble to a computation—if possible, practical—in group theory involving just the Nielsen classes an the action of the Hurwitz monodromy group.

In §3.2 we give the computational approach to the action of a subgroup $H_{\sigma}$ of $H(r)$ c the fundamental group $\pi_1(X)$ (or on $H_1(X, \mathbf{Z})$) of a cover $X \to \mathbf{P}^1_x$ in a given absolute Nielse class $\mathrm{Ni}(\mathbf{C})^{ab}_T$. The subgroup $H_{\sigma}$ is the stabilizer in $H(r)$ of an element $\sigma \in \mathrm{Ni}(\mathbf{C})^{ab}_T$. Since th note is intended to be expositional we comment on just two points: for $g = 1$ or 2, the actic of $H_{\sigma}$ on $H_1(X, \mathbf{Z})$ is through a finite group if and only if the moduli dimension of the Nielse class is 0; and the general computation of whether the action is through a finite group must b difficult (albeit, primitive recursive). Finally, we illustrate the "endomorphism computation discussed in [FrG; §5] and §2.3 by mentioning the problems in computing those endomorphism of $H_1(X, \mathbf{Q})$ that arise from the group ring $\mathbf{Z}[G]$ in this special case: $G = A_n$; and the cover is i the Nielsen class $\mathrm{Ni}(\mathbf{C})$ with each of the conjugacy classes in $\mathbf{C}$ equal to the conjugacy class of 3-cycle. The main tool here is just the Lefschetz trace formula.

## §2. VOL. 3 OF ZARISKI'S COLLECTED WORKS APPLIED TO [FrG]

The papers of [Z] that apply to this note and to [FrG] are as follows:

- [8] Sull'impossibilitá di risolvere parametricamente per radicali un'equazione algebrica $f(x, y) = 0$ di genere $p > 6$ a moduli generali, 1926, 43–49.
- [12] Sopra una classe di equazioni algebriche contenenti linearmente un parametro e risolub per radicali, 1926, 58–80.
- [13] On a theorem of Severi, 1928, 81–86.
- [18] On the moduli of algebraic functions possessing given monodromie group, 1930, 155–17
- [28] On the Poincaré group of rational plane curves, 1936, 266–278.
- [29] A theorem on the Poincaré group of an algebraic hypersurface, 1937, 279–289.
- [31] The topological discriminant group of a Riemann surface of genus $p$, 1937, 307–330.

There are comments by M. Artin and B. Mazur [AM] on [8] and [12] on p.2, [13] on p.3– and on [28], [29] and [31] on p.9–10. There are no comments on [18], which does happen to be the center of our discussion. What we call the Hurwitz monodromy group, $H(r)$ in §1.2 appea in [28] where Zariski says (p.266) that it "practically coincides with the braid group." Also, [AM p.9] say that this is essentially the $r$th braid group defined by Artin. There is, however, a practic distinction between the groups that requires more than a "Hey, you!" when it is the turn of $H(r$ No one has yet personally objected to the author's naming of $H(r)$. So it shall stand with th author until good objection comes forth. It was M. Artin (in 1972 during the writing of [Fr,1 who pointed out the prior application that Hurwitz [Hu] had made of it to simple branched cover An aside: It is traditional in classical Riemann surface theory to use $p$, instead of $g$, for the genu

## §2.1. ON [8] AND THE FOUNDATIONAL THEOREM OF [FrG].

Zariski's papers are "talky" in comparison to modern papers in algebraic geometry. Some cautic s advisable since several of them are in Italian. Curiously, the preprint title of [FrG], prior to our wareness of Zariski's work was essentially the English version of the title of [8].

**tatement 2.1:** *[8] contains the proof that the generic curve of genus $g > 6$ has no map to $\mathbf{P}^1_x$ vith solvable monodromy group.*

he old Italian understanding of the word generic was quite loose, but it definitely is used here s in [FrG] and as in the notation of §1.1. Consider a group G with a faithful permutation epresentation $T : G \to S_n$. Let $\mathcal{M}_g(G)$ be the collection of points $\boldsymbol{m} \in \mathcal{M}_g$ such that $\boldsymbol{m}$ s represented by a curve $X_{\boldsymbol{m}}$ that has a map $\phi : X \to \mathbf{P}^1_x$ with monodromy group G (and ermutation representation—via the cover—equal to $T$). If $G$ runs over solvable groups and $> 6$, then $\mathcal{M}_g(\text{sol}) = \bigcup_G \mathcal{M}_g(G)$ contains no nonempty Zariski open subset of $\mathcal{M}_g$. This is nuch the same as the main part of the statement of [FrG; Prop. 3.1].

**tatement 2.2:** *Historical background.*

ariski attributes the main lemma for the reduction to the primitive case to the 1897 International ongress talk of Enriques. Enriques also states, as unsolved, the problem of showing that the eneric curve of genus $g > 6$ is not uniformized by radicals. The reduction to the primitive case a [FrG] comes from a throw-away paragraph in [Fr,1; p.26], but it was for exposition purposes here—no claim was made of originality. The motivation for consideration of the problem by [FrG] s manifold, and it has been pushed forth at this time as a part Thompson's program (Statement .6). More modest motivation comes from [FrJ; p.137] which has a near outline of the proof of rop. 3.1 of [FrG]; and it notes that if this were false, then most likely the solvable closure of he rationals would be a PAC (pseudo- algebraically closed) field. This latter question is still ndecided. Actually (and Mumford noted this independently), there is an apparently much more ifficult question: Given $g$, is the function field of the general curve of genus $g$ a subfield of the unction field of some $X$ where $X \to \mathbf{P}^1_x$ has solvable monodromy group. If someone doesn't show his is impossible for large $g$ soon, [FrG] will comment on why we believe this is difficult.

**tatement 2.3:** *Zariski on the exceptional values of $g$.*

ere we take $g$ to be exceptional if $\mathcal{G}_g(\text{sol})$ is nonempty. Zariski is satisfied to comment on the xceptional values of g by noting that $\mathcal{G}_g(\text{prim})$ contains $S_{[g+3/2]}$ which is a solvable group in ach of the cases $g$=1,...,6. Artin-Mazur comments [AM;p.2], using the language from classical iemann surface theory, that the existence of a $g^1_4$ in the case of each curve of genus 6 was still nproved at the time of Zariski's paper because of gaps in papers of the Italian school. The use f [KL] in [FrG] is exactly what Artin-Mazur recommend, and therefore [FrG; Principle 2.5] is ght on target in declaring that $S_n \in \mathcal{G}_g(\text{prim})$ if and only if $n \geq [g + 3/2]$. This, of course, is oundational for the interesting calculations that have arisen in [FrG; §5.3] for deciding, given $g$, nose $n$ for which $A_n \in \mathcal{G}_g(\text{prim})$.

## 2.2. [12]; EXCEPTIONAL SOLVABLE GROUPS; THOMPSON'S PROGRAM

**tatement 2.4:** *Zariski's use of group theory and Ritt's results.*

he argument used by Zariski [8; p.46-49] to conclude his equivalent to [FrG; Prop. 3.1] is slightly nger than the one page of [FrG], and it is less valuable for listing possible exceptions. It sts, however, on exactly the same group theoretical principles—going back to Galois. This is minuscule portion of [FrG]. Justification for the the work of [FrG; §3 and §4] comes from the esire to display all "branch cycles" $\boldsymbol{\sigma} = (\sigma_1, \ldots, \sigma_r)$ for all of the solvable groups $G$ $(= G(\boldsymbol{\sigma}))$ ith these properties:

$$\sum_{i=1}^{r} \text{ind}(\sigma_i) = 2(n + g - 1) \quad \text{and} \quad r \geq 3g \,, \tag{2.1}$$

here g is the genus of a cover $X \to \mathbf{P}^1_x$ of degree $n$ with $\boldsymbol{\sigma}$ as a description of its branch cycles.

Zariski's own papers provide two motivations for this. The most important of these is [18], which is also the topic of [FrG; §5]. Statement 2.8 reports that his conjectures in [18] a wrong—by examples that already appear in [FrG; §5.2] (e.g., following Ries [R]). An explanatic of the condition $r \geq 3g$ appears in §2.3.

Here, however, he was already ahead of his time. More naive motivation appears in [12 based on work of Chisini and Ritt [Ri].

**Statement 2.5:** *How Zariski continues a long tradition.*

In his introduction Zariski mentions papers of Klein (1874), Bianchi and Chisini (1900) and Chisi (1915) as precursers to Ritt's classification of the covers $\phi : X \to \mathbf{P}^1_x$ with the genus $g(X)$ of equal to 0, deg($\phi$) a prime and with solvable monodromy group. Ritt observes that the Galo closures $\hat{X}$ of each of his 5 cases are of genus 0 or 1. This puts an interesting structure into th problem. Although the Schur problem for rational functions [Fr,2; p.148] seems to have nothin to do with "uniformization by radicals," after a quote of a theorem of Burnside the geometr territory for the Schur problem of prime degree is the same as given by Ritt's list [12; p.59]. was precisely the availability of the *arithmetic* of elliptic curves that solved the Schur proble for rational functions of prime degree. We quote [AM; p.2]:

> "... Zariski generalizes their [Chisini and Ritt] elegant result to arbitrary degree. He classifies solvable [$\phi : \mathbf{P}^1_y \to \mathbf{P}^1_x$] having the familiar property that all points of [$\phi^{-1}(x)$] can be expressed rationally in terms of any pair. The key point is that the [Galois closure of $\mathbf{P}^1_y \to \mathbf{P}^1_x$] is either rational or an elliptic curve."

Despite our own personal motivations just mentioned, and in light of the rather obvio group theory, it is hard for us to understand why Artin and Mazur make much of the Chisini-Ri result considering what they ignore. John Thompson suggests that an error in Galois [B; p.16 165] persisted as the motivation for the problem. Galois incorrectly asserted that it was usu for solvable covers to have the property of the quote above. (He knew of couterexamples, bu considered them, contrary to modern understanding, to be rare.) Zariski points out that his cla sification yields the expected generalizations excluding only one new type—"scarcely of interest (we agree)—from a cover of degree 4. The proof is long and we know of no unusual applicatior or reasonable generalizations.

**Statement 2.6:** *The geometric portion of Thompson's program and moduli dimension.*

There are two main conjectures about the monodromy group of covers in Thompson's prograr

**Solvable group conjecture:** *Excluding the Ritt-Zariski list there are only finitely many prim tive solvable groups that occur as monodromy groups of covers $X \to \mathbf{P}^1_x$; and for each $g \geq 1$ the are only finitely many solvable primitive groups that occur as the monodromy groups of cove $X \to \mathbf{P}^1_x$ with $g(X) = g$.*

Actually the case $g = 0$ is quite significant, but it is already a theorem [GTh]. Clear $\mathcal{G}_0$(sol) is entirely composed of subgroups of sequences of wreath products from $\mathcal{G}_0$(prim)$\cap\mathcal{G}_0$(so (§1.1) and a major portion of the groups of $\mathcal{G}_g$(sol) are comprised from subgroups of wreat products of elements of $\mathcal{G}_g$(prim)$\cap\mathcal{G}_g$(sol) and $\mathcal{G}_0$. A brave venture might be that in the "Solvab group conjecture," excluding the Ritt-Zariski list, $A_n$ and $S_n$, $n = 5, 6, \ldots$—these both occur f several distinct types of Nielsen classes even in genus 0—there are only finitely many primiti groups that occur as monodromy groups of covers $X \to \mathbf{P}^1_x$ with $g(X) = g$. The followin however, seems more certain.

**Composition factor conjecture:** *Excluding $A_n$ and cyclic groups, for each $g \geq 0$ there a only finitely many simple groups that occur as composition factors of monodromy groups of cove $X \to \mathbf{P}^1_x$ with $g(X) = g$.*

Ultimately the program is concerned with the arithmetic of these covers, but it is easier state the results of [FrG] and [GTh] if we bring up just one geometric quantity, *moduli dimensio* the moduli dimension of $(g, G)$ is the dimension of $\mathcal{M}_g(G)$ (Definition 1.6). For $g > 1$ the moduli dimension of $(g, G)$ is at most $3g - 3$. Note that we assume that $G$ is a transitive permutation group, and that the permutation representation of the monodromy group of the cover is the same as that attached to the group. When the the moduli dimension is maximal possible, Zariski says that $G$ is *nonspecial*. Even though Zariski got some motivation from the appearance of *special* divisors in [18] (c.f. Statement 2.7) the word seems bland. Therefore we say that $(g, G)$ has *full moduli dimension*. In the context of [FrG], when there can be no confusion we might say that $(g, G)$ (or when $g$ is understood, just $G$) is *exceptional*.

## §2.3. ON [18]:MODULI DIMENSION FOR THE EXCEPTIONAL $g$

**Statement 2.7:** *Groups with sufficient branch point parameters to have a chance to have full moduli dimension.*

Since the moduli space of curves of genus $g$ is of dimension $3g - 3$ it is an ancient observation, based on Principle 1.6, that in order that $(g, G)$ have full moduli dimension there must be a cover $\phi : X \to \mathbf{P}^1_x$ with monodromy group $G$, $g(X) = g$ and at least $r = 3g$ branch points. It is obvious that this is not (usually) sufficient unless the group is primitive—equivalently, no curves are properly contained between $X$ and $\mathbf{P}^1_x$. Zariski notes this by example. Then he repeats several times [18; p.156,157 and generally along into the paper] that he believes that if $r \geq 3g$ and $G$ is primitive, $G$ has full moduli dimension. He phrases the remainder of his paper as generalization of the result of [KL] and he develops a complicated formula for the moduli dimension.

Well, actually, it isn't really so complicated from a modern viewpoint, for he has actually rephrased the problem in terms of the dimensions of fibers of the Picard bundle

$$\underbrace{X \times X \times \cdots \times X}_{n \text{ times}} / S_n \to \mathrm{Pic}^n(X), \tag{2.2}$$

where $X \times X \times \cdots X/S_n \stackrel{\mathrm{def}}{=} X^{(n)}$ denotes the symmetric product of $X$, $n$ times, and $\mathrm{Pic}^n(X)$ denotes the divisor classes of degree $n$ on $X$, as $X$ varies over representatives of points of a Zariski open subset of $\mathcal{M}_g$.

In particular [FrG] provides a list of test examples for the solvable group version of the problem of computing the moduli dimension of $(g, G)$. If we exclude $S_n$'s then 0,1 and 2 are the only values of $g$ for which $(g, G)$ has full moduli dimension with $G$ a (primitive) solvable group. But, in the case $g = 2$ the four groups that are monodromy groups of covers with at least 6 branch points [FrG; Theorem 3.3] appear in the statement of Theorem 1.2.

Indeed, [FrG; §4] lists the complete set of *Nielsen classes* that are associated to covers with these groups as monodromy groups, and thus through Hurwitz monodromy action collects the covers into algebraic subsets whose images in $\mathcal{M}_g$ will be irreducible varieties. Much of [FrG; §5] brings together tools that bear on which of these examples do have full moduli dimension. Contrary to Zariski's conjecture: Not all (Statement 2.8)!

The list of the cases where $g = 1$, $G$ is primitive solvable $G$ and the Nielsen class has full moduli dimension (=1) is not complete. All have degree $p^r$ with $p \leq 7$ [FrG; Theorem 3.2] and it is likely that the exceptions include the list with g=2. But this hasn't yet been checked. The list for g=0 (also not yet complete) is part of [GTh], and it includes many groups that don't appear in the list for $g = 2$.

**Statement 2.8:** *The endomorphism test for complete moduli dimension in the case $g = 2$ gives counterexamples to the main conjecture of [18].*

Suppose we are given a cover $\phi : X \to \mathbf{P}^1_x$. Let $\hat{\phi} : \hat{X} \to \mathbf{P}^1_x$ be the Galois closure of this cover. There is a fairly explicit algorithm for computing those endomorphisms of the Jacobian, $J(X)$, of X (identified with $\mathrm{Pic}^0(X)$) that arise from the group ring $\mathbf{Z}[G]$ with $G = G(\hat{X}/\mathbf{P}^1_x)$ acting on $J(\hat{X})$ (leaving $J(X)$ stable). Indeed, in the examples here, and in the other test case $(g, A_n)$ of

[FrG; §5.3], it suffices to carry out all computations on $H_1(X, \mathbf{Z})$ using just the Lefschetz trace formula ([FrG; Theorem 5.5] and §3.2).

Identify $H_1(X, \mathbf{Q})$ as the image of $H_1(\hat{X}, \mathbf{Q})$ under $pr_X = \sum_{\sigma \in G(\hat{X}/X)} \sigma$ and consider those elements of $\mathbf{Q}[G]$ that commute with $pr_X$. Denote the action of this subring on $H_1(X, \mathbf{Q})$ by $\mathrm{End}_{\hat{X}}(X)$. If $\mathrm{End}_{\hat{X}}(X)$ properly contains $\mathbf{Q}$, then $(g, G)$ doesn't have full moduli dimension (c.f. Comments on [13]). These computations aren't yet complete, but two at least of the exceptional groups in the case of g=2 don't have full moduli dimension by this criteria, and none of the list of (1.3) yet has been shown to have full moduli dimension. Ries was the first to give an example of the failure of Zariski's conjecture [R] (for $D_{10}$—albeit in a somewhat intricate format).

**Statement 2.9:** *Continuation of $g = 2$, a full moduli formula and the hyperelliptic involution.*

The spaces $\mathcal{M}_g$ are special for the values $g = 1$ and 2. They are *affine* open subsets of a natural (Igusa) projective compactification. There is a general test for moduli dimension being 0 that is "if and only if" in the case $g = 1$ or 2: All of the Picard-Lefschetz transformations around "branches at $\infty$" act on $H_1(X, \mathbf{Q})$ as elements of finite order. By identifying the "braid group" generators of the Hurwitz monodromy group with P-L transformations and by expressing $H_1(X, \mathbf{Z})$ in terms of "branch cycles," ([FrG; Theorem 5.4] and §3, Theorem 3.6) provides lower bounds for the moduli dimension. It would be a considerable refinement of the ideas attributed to Mayer-Mumford in [Gr; §13] to rephrase the whole problem of computing the moduli dimension in terms of P-L transformations even in Hurwitz family situations where, as we have just noted, things can be computed explicitly. Also, "coalescing" of branch cycles provides much information on boundary behaviour of the Hurwitz family, and indirectly on the moduli dimension.

In the case $g = 2$, when it is possible to reconstruct the canonical involution from the branch cycles in the Nielsen class giving the collection of covers, then we have a precise handle (via Igusa) on the relation of the family of covers to $\mathcal{M}_g$. We could do this from pure group theory if we are in a situation where our cover $\phi : X \to \mathbf{P}^1_x$ is part of a commutative diagram

$$\text{(2.4)} \qquad \begin{array}{ccccccc} \hat{X}_u & \longrightarrow & \hat{X}_x & \longrightarrow & X & \xrightarrow{\lambda} & \mathbf{P}^1_t \\ & & & & \downarrow{\scriptstyle \phi} & & \downarrow{\scriptstyle \phi(h)} \\ & & & & \mathbf{P}^1_x & \xrightarrow{\phi(f)} & \mathbf{P}^1_u \end{array}$$

where we have renamed $\hat{X}$ to be $\hat{X}_x$, $\lambda : X \to \mathbf{P}^1_t$ is the canonical hyperelliptic involution and $\hat{X}_u$ is the Galois closure of the cover $X \to \mathbf{P}^1_u$. The maps $\phi(f)$ and $\phi(h)$ derive from rational functions f and h.

Of course, the most important point of this diagram is that there are such rational functions. This is not to be expected in general. But we don't know if this is precluded for Nielsen classes of full moduli dimension. It is an example of the "finite correspondence situation" that was featured in [Fr,3]: given everything in the diagram, except the lower right corner, we ask when we can fill it in using rational function maps f and h.

**The Basic Problem:** *Which covers $\phi : X \to \mathbf{P}^1_x$, expressed in terms of branch cycles are part of a diagram like (2.4).*

**Comments:** Since the cover $\phi : X \to \mathbf{P}^1_x$ is primitive, we may assume that $X$ is a component of the fiber product $\mathbf{P}^1_x \times_{\mathbf{P}^1_u} \mathbf{P}^1_t$, and the maps $\phi$ and $\lambda$ arise from projection on the two factors. The Galois closures of each of these covers give groups, $G_f$ and $G_h$, that are homomorphic images of the group $G_u = G$. This can be rephrased entirely in terms of group theory with "branch cycle" generators $\boldsymbol{\tau}$.

Assume given a description $\boldsymbol{\sigma}$ of the branch cycles of $\phi : X \to \mathbf{P}^1_x$. We seek a group $G_u$ with "branch cycle" generators $\boldsymbol{\tau}$ and with three transitive permutation representations (named for the above situation) $T_X$, $T_f$ and $T_h$ with these properties:

(2.5) a) $T_X$ arises from an orbit of $T_f \otimes T_h$;

b) $\boldsymbol{\sigma}$ is in the Nielsen class of a Schreier construction arising from $T_f(\boldsymbol{\tau})$ applied to $G(T_X)$, the stabilizer of an integer in the representation $T_X$; and

c) branch cycles for a degree 2 hyperelliptic cover are the result of a Schreier construction arising from $T_h(\boldsymbol{\tau})$ applied to $G(T_X)$.

We don't explain the phrase "Schreier construction", but it is essentially the construction free generators of a subgroup of finite index in a free group. Without experimenting it is difficult say whether, given $\boldsymbol{\sigma}$, the existence of $G_{\boldsymbol{u}}$, $\boldsymbol{\tau}$, etc. is a reasonably effective calculation, but it esn't look easy. □

## .4. ON [13], [28]–[31]: ENDOMORPHISMS ARE SPECIAL AND MORE ON $H(r)$

**atement 2.10:** *Comments of [AM; p.3–4] on Endomorphisms of generic curves.*

t $C$ be a curve and $T$ a curve on $C \times C$. Then $T$ is called a singular correspondence if its ıomology class in $H^2(C \times C, \mathbf{Q})$ isn't a linear combination of the classes of the diagonal, $\boldsymbol{p} \times C$ d $C \times \boldsymbol{p}$ with $\boldsymbol{p} \in C$. Hurwitz conjectured and Severi tried to prove that a sufficiently general ear system $|C|$ on a surface $F$ contains no singular correspondence. He based his argument on nilies of plane curves with only nodes as singularities. As a consequence the result of Statement follows.

Zariski points out that there are difficulties with Severi's argument, but concludes the ult in the case that the rational map from $F$ to $\mathbf{P}^1_x$ induced from $|C|$ is birational. In the proof uses a result of Severi that has been proved only for Lefschetz pencils. Here [AM; p.4] claim ıt it is not difficult to verify this extra condition in the case of plane curves with nodes—thus result: "So the proof that the general curve of genus [$g$] has no singular correspondence can distilled from the two papers of Severi and Zariski."

Also [AM] includes an allusion to a *preprint* of Mori, but there is no description of ıtents, so its relation to the topic of correspondences is obscure. What was left out was any erence to Lefschetz's paper [L] which appeared in the same volume as Zariski's, even though riski himself includes it as a footnote. Lefschetz is quite clear: He shows that if the "abelian ctions of every [curve of genus g] have a complex multiplication, then there exists a fixed nplex multiplication common to them all;" and then he shows, by an explicit induction on -using explicit computation of periods—that the general hyperelliptic curve has no complex ltiplication.

**atement 2.12:** *Zariski seems to be the first to write out the relation between the Artin braid up and the Hurwitz monodromy group.*

iski's interest in the Hurwitz monodromy group seems to have nothing to do with moduli ıilies. Indeed, the only evidence in [Z] of any motivation coming from classical moduli space nking related to families of curves is his considerable work on curves in $\mathbf{P}^2$ with, say, only les (or nodes and cusps) as singularities. But even here he concentrates on the fundamental up of the complement of such a curve. Here is the progression of his papers on this.

In [29] he argues for $r > 2$ that if $V^{r-1}$ is a hypersurface in $\mathbf{P}^r$, then $\pi_1(\mathbf{P}^r - V^{r-1})$ is norphic to $\pi_1(L - L \cap V^{r-1})$ where $L$ is a generic hyperplane section of $\mathbf{P}^r$. There is a claim AM; p.7] that his proof requires amplification on several points. They mention [AM; p.15] the rse theory proof of D. Cheniot.

In [28] Zariski notices that "maximal cuspidal curves" of even order $2r-2$ are generic ions of the discriminant locus, and that among rational curves $C$ in $\mathbf{P}^2$ with only nodes and ps, other than the maximal cuspidal curves (recall the connection with the Hurwitz monodromy up), $\pi_1(\mathbf{P}^2 - C)$ is cyclic.

Finally, in [31] there is something that the author hasn't seen used before. Use the ation of (2.2) with $D_r$ denoting the natural discriminant locus in $X^{(r)}$. For a nonsingular jective curve $X$ of genus g, not only does Zariski compute $\pi_1(X^{(r)} - D_r)$, denoted $G_{r,g}$, in ns of generators and relations, but he considers a fascinating normal subgroup of it. Assume $r > g$ and apply the map (2.2) from $X^{(r)}$ to $\mathrm{Pic}^r(X)$. The general fibers of this are well wn to be copies of $\mathbf{P}^{r-g}$. Thus, in Zariski's language, $X^{(r)}$ contains a system of $\infty^g$ , $\mathbf{P}^{r-g}$'s.

He then computes $\pi_1(\mathbf{P}^{r-g} - D')$ where $D'$ is the intersection of the discriminant locus wit general one of these $\mathbf{P}^{r-g}$'s.

## §2.5. THE SOLVABLE AND GENUS 0 HULLS

**Statement 2.13:** *The genus zero hull of an element of a function field.*

The title contains a neologism which we will define as follows. Let $x$ be a nonconstant funct on a curve $X$ (over $\mathcal{C}$). Denote an algebraic closure of $\mathcal{C}(X)$ by $\bar{K}$. Starting from $x$ consider smallest field $\bar{K}_o$ containing $\mathcal{C}(x)$ that includes the Galois closure of each field extension $\mathcal{C}(u)/\mathcal{C}$ with nonconstant $u$ and $v \in \bar{K}_o$. Part of Thompson's program is to describe $\bar{K}_o$ (and the simila defined $\bar{K}_g$ where $\mathcal{C}(u)$ is replaced by any function field of genus $g$). Note that this includes d on the lattice of subfields of $\mathcal{C}(x)$ compatible with the discussion of Statement 2.9 and of ma of Ritt's papers (cf. [Fr,3]).

**Statement 2.14:** *For which $g$ is $\mathcal{M}_g(\text{sol})$ $(= \cup_{G \text{ solvable}} \mathcal{M}_g(G)$ (conclusion of §1.1 and Sta ment 2.1) dense in $\mathcal{M}_g$?*

The result of Zariski does not preclude that the $g$'s that satisfy this statement exceed 6. E in this direction it seems possible that for fixed $g > 0$ there exists $N = N(g)$ such that $n > N$ there are no primitive solvable groups of genus $g$ (i.e., appearing as monodromy grou of covers of $\mathbf{P}^1_x$ by some genus g curve—c.f. Statement 2.6). Recall that $n$ denotes the degree the permutation representation that goes with $G$. Some understanding of the solvable closure $\mathcal{C}(x)$ would follow from this if we also understood, for fixed $g$, how to bound the degrees of covers $X \to Y$ with solvable monodromy group where $g(X) = g$ and $g(Y) > 0$.

**Statement 2.15:** *Nielsen classes consisting entirely of hyperelliptic curves, and generalities lating one Nielsen class to another.*

Suppose given a Nielsen class Ni(**C**) where **C** represents an r-tuple of conjugacy classes in group G. Statement 2.9 can be generalized beyond the case $g = 2$. We can inspect whetl the set of covers $\phi : X \to \mathbf{P}^1_x$ in this Nielsen class each fit in a commutative diagram like (2 that displays a hyperelliptic involution for $X$. This implies the existence of (complex analyt $\Psi(\mathbf{C}, \mathbf{C}') : \mathcal{H}(\mathbf{C}) \to \mathcal{H}(\mathbf{C}')$ where $\mathbf{C}'$ denotes the Nielsen class for hyperelliptic covers of t genus of the family. But such a morphism exists if each $X$ that appears in the Nielsen cl is hyperelliptic. If this situation occurs, we would have all of the apparatus for computing t moduli dimension of the Nielsen that arises from the special "Igusa-like" compactification of t hyperelliptic curves of genus $g$.

More generally, for a given Nielsen class Ni(**C**) we would ask how one might effectiv compute the possibility for a natural map $\Psi(\mathbf{C}, \mathbf{C}') : \mathcal{H}(\mathbf{C}) \to \mathcal{H}(\mathbf{C}')$ where $\mathbf{C}'$ denotes so other Nielsen class. We shall say that **C** and $\mathbf{C}'$ are *concatenated* if the obvious analogue diagram (2.4) exists. It must be a difficult problem to decide if **C** and $\mathbf{C}'$ are concatenated. T main theorem of [ArC] considers the case that the conjugacy classes of the Nielsen class **C** are 2-cycles and the degree of the representation is smaller than $g/2 + 1$. If the degree of the gro associated to C$'$ also does not exceed $g/2+1$ then any concatenation must be particularly simp the equivalence classes of the covers represented by points of $\mathcal{H}(\mathbf{C}')$ are of the form $X \to \mathbf{P}^1_x \to$ where the covers $X \to \mathbf{P}^1_x$ are in Ni(**C**).

**Statement 2.16:** *The relation between Nielsen classes for a subgroup $H$ of $G$ and Nielsen clas for $G$.*

Suppose that Ni(**C**) is a Nielsen class for the group $H \subset G$, and assume that Ni(**C**) has full mod dimension. In the light of Principle 1.6 it is tempting to think that there must be a Nielsen cl Ni($\mathbf{C}'$) for $G$ for which Ni($\mathbf{C}'$) is also of full moduli dimension. But the principle of "coalesci of branch cycles" would say that there is probably little chance that such a statement hol generally without adding the the following condition; in which case the statement is true by t same ideas that give Principle 1.6. Suppose that $\boldsymbol{\sigma} \in \mathrm{Ni}(\mathbf{C})$. There should exist generators of $G$,

$$\sigma_{1,1}, \ldots, \sigma_{1,s_1}, \sigma_{2,1}, \ldots, \sigma_{2,s_2}, \ldots, \sigma_{r,1}, \ldots, \sigma_{r,s_r}, \ i = 1, \ldots, r,$$

whose sum of indices gives the same genus $g$ as does $\boldsymbol{\sigma}$ and with $\sigma_{i,1} \cdots \sigma_{i,s_i} = \sigma_i$, $i = 1, \ldots, r$. In practical situations checking for this situation would be a nontrivial computation.

## §3. HURWITZ MONODROMY ACTION ON $\pi_1(X)$

Use the notation of the previous sections for an absolute Nielsen class $\mathrm{Ni}(\mathbf{C})^{ab}_T$. Consider the Hurwitz space $\mathcal{H}(\mathbf{C})_T$ parametrizing equivalence classes of covers $X \to \mathbf{P}^1_z$ in this Nielsen class. We assume that this representative cover corresponds to a point $\boldsymbol{m}_0 \in \mathcal{H}(\mathbf{C})_T$. The goal of this section is to discuss the combinatorial computation of the natural action of the fundamental group of $\mathcal{H}(\mathbf{C})_T$ on the fundamental group $\pi_1(X)$. This starts with the computation of $\pi_1(X)$ in terms of branch cycles for the cover $X \to \mathbf{P}^1_z$ (§3.1, Theorem 3.5). As usual $g = g(X)$ is the genus of $X$.

The action of $\pi_1(\mathcal{H}(\mathbf{C})_T)$ on this (§3.2) can be regarded as giving data for a global version of a local problem, the computation of the variation of the complex structure of $X$ as $\boldsymbol{m}$ varies in a suitably small neighborhood $U$ of $\boldsymbol{m}_0$ on $\mathcal{H}(\mathbf{C})_T$. This local problem can be made quite explicit by considering an analytic basis $(\omega(\boldsymbol{m})_1, \ldots, \omega(\boldsymbol{m})_g)$ for the space of holomorphic 1-forms $\Gamma(\Omega^1(X_{\boldsymbol{m}}))$ on $X_{\boldsymbol{m}}$ as $\phi_{\boldsymbol{m}} : X_{\boldsymbol{m}} \to \mathbf{P}^1_z$ runs over representatives for the equivalence class of covers $\boldsymbol{m} \in U$. Denote the tangent space to $\mathcal{H}(\mathbf{C})_T$ at $\boldsymbol{m}_0$ by $\mathbf{T}_{\boldsymbol{m}_0}$. The directional derivatives $D_{\boldsymbol{t}}$ applied to $(\omega(\boldsymbol{m})_1, \ldots, \omega(\boldsymbol{m})_g)$, $\boldsymbol{t} \in \mathbf{T}_{\boldsymbol{m}_0}$—the result is a vector of differentials of second kind at worst—gives sufficient information to determine the moduli dimension of the Nielsen class. This is complicated, but it is outlined in the opening section of [Gr] under the heading of the Gauss-Manin connection. A simple case is illustrative.

Since we may choose $U$ so that the family of covers over $U$ is locally constant we may regard $H^1(X_{\boldsymbol{m}}, \mathcal{C})$ as constant in $\boldsymbol{m}$. Suppose that the application of $D_{\boldsymbol{t}}$ to $(\omega(\boldsymbol{m})_1, \ldots, \omega(\boldsymbol{m})_g)$ gives an analytic vector of *holomorphic* differentials on $X_{\boldsymbol{m}_0}$ for each $\boldsymbol{t} \in \mathbf{T}_{\boldsymbol{m}_0}$. Then the vector subspace $\Gamma(\Omega^1(X_{\boldsymbol{m}}))$ is a constant subspace of $H^1(X_{\boldsymbol{m}}, \mathcal{C})$ as $\boldsymbol{m}$ varies. From the local Torelli theorem [Gr; p.247] the moduli dimension of the Nielsen class is 0.

The abstract construction of the Hurwitz family would seem to mitigate against the computation of the holomorphic differentials $(\omega(\boldsymbol{m})_1, \ldots, \omega(\boldsymbol{m})_g)$. Data from Nielsen classes—of an essentially global nature—gives the appearance of a more tractible computation. Since, however, the theory for determining the moduli dimension from such data hasn't been completed, we make a local–global comparison here only for when the moduli dimension is nonzero.

### §3.1. FUNDAMENTAL GROUP COMPUTATION FROM BRANCH CYCLES

Given a cover $\phi : X \to \mathbf{P}^1_z$ we compute $\pi_1(X)$ (and $H_1(X, \mathbf{Z})$) in terms of a description of the branch cycles for the cover (Theorem 3.5). Indeed, if the cover is Galois it is back to the well known Schreier construction for generators of a subgroup of a free group in an explicit way.

**The Galois case:** Denote by $S = \{\bar{\sigma}_1, \ldots, \bar{\sigma}_r\}$ generators of the free group $F_r$ on $r$ generators. With $\boldsymbol{\sigma}$ a description of the branch cycles of the cover, let $G$ be $G(\boldsymbol{\sigma})$ and $G(1) = \{\gamma \in G(\boldsymbol{\sigma}) \mid \gamma(1) = 1\}$. Consider the homomorphism $\delta : F_r \to G(\boldsymbol{\sigma})$ induced by $\bar{\sigma}_i \to \sigma_i$, $i = 1, \ldots, r$. Finally, let $H(1)$ be $\delta^{-1}(G(1))$.

In order to get free generators of $H(1)$ we need a function $\rho : F_r \to F_r$ representing right cosets of $H(1)$, with the following properties : $\rho(1) = 1$ ; $\rho(\alpha) \in H(1)\alpha$ ; and $\rho(h\alpha) = \rho(\alpha)$ for each $h \in H(1)$ and $\alpha \in F_r$. Furthermore $\rho$ may be selected to have the following property:

$$\text{(3.1)} \qquad \mathrm{length}_{\bar{\sigma}}(\rho(\alpha)) = \min_{h \in H(1)} \mathrm{length}_{\bar{\sigma}}(h\alpha) \quad \text{for} \quad \alpha \in F_r$$

where $\mathrm{length}_{\bar{\sigma}}$ denotes the length of a word in the $\bar{\sigma}$'s.

Automatic from this is the following property: if $\rho(\alpha) = s_1^{\epsilon_1} \cdots s_n^{\epsilon_n}$ is a reduced representation of $\rho(\alpha)$, $s_i \in S$, $\epsilon_i \in \{\pm 1\}$, $i = 1, \dots, n$, then

$$s_1^{\epsilon_1} \cdots s_i^{\epsilon_i} \in \rho(F_r) \quad \text{for each } i = 1, \dots, n. \tag{3.2}$$

With these conditions, the collection

$$M = \{rs\rho(rs)^{-1} \mid r \in \rho(F_r),\ s \in S \text{ and } rs \notin \rho(F_r)\}$$

generates H(1) freely (e.g., [ FrJ; Lemma 15.23 ] ).

**Lemma 3.1:** *If $\phi : X \to \mathbf{P}^1_x$ is a Galois cover, then the fundamental group of $X$ is isomorphic to the image of $H(1)$ in $F_r/N$ where N is the smallest normal subgroup of $F_r$ containing $\bar\sigma_1 \cdots \bar\sigma_r$ and $\bar\sigma_i^{\mathrm{ord}(\sigma_i)}$, $i = 1, \dots, r$.*

This definitely doesn't hold if $\phi : X \to \mathbf{P}^1_x$ isn't Galois . Here is the easiest example.

**Example 3.2:** *$S_3$ in its regular and nonregular representations.* Let $G$ be $S_3$, let $r > 3$ be an even integer and let $\mathrm{C}_i$ be the conjugacy class of a 2-cycle in $G$, $i = 1, \dots, r$. Let $\phi : X \to \mathbf{P}^1_x$ be a cover in the Nielsen class Ni(**C**) where $G$ is in its representation of degree 3. As usual let $\hat\phi : \hat X \to \mathbf{P}^1_x$ be the Galois closure of $\phi : X \to \mathbf{P}^1_x$. Then we may compute $\pi_1(\hat X)$ from Lemma 3.1. For simplicity, and with no loss, assume that $\sigma_1 = \sigma_2 = (1\,3)$, and that $\sigma_3 = \cdots = \sigma_r = (1\,2)$. Then $\{1, \bar\sigma_2, \bar\sigma_3\} = \Lambda$ consists of coset representatives for $H(1)$, and

$$M = \{\bar\sigma_1\bar\sigma_i,\ \bar\sigma_3\bar\sigma_i\bar\sigma_3^{-1},\ i = 1,2,\ \bar\sigma_3\bar\sigma_j,\ \bar\sigma_1\bar\sigma_j\bar\sigma_1^{-1},\ j = 3, \dots, r,\ \bar\sigma_2\bar\sigma_1^{-1},\ \bar\sigma_j\bar\sigma_3^{-1},\ j = 4, \dots, r\}$$

generates $H(1)$ freely.

Similarly , appropriate coset representatives for $\widehat{H(1)}$ defined by the regular representation of $G$ are given by $\hat\Lambda = \Lambda \cup \{\bar\sigma_1\bar\sigma_3,\ \bar\sigma_3\bar\sigma_1,\ \bar\sigma_1\bar\sigma_3\bar\sigma_1\}$. Thus,

$$\begin{aligned}\hat M =\{&\bar\sigma_2\bar\sigma_1^{-1},\ \bar\sigma_j\bar\sigma_3^{-1},\ \bar\sigma_1\bar\sigma_j(\bar\sigma_1\bar\sigma_3)^{-1},\ j = 4, \dots, r,\ i = 1,2,\\ &\bar\sigma_3\bar\sigma_j,\ j = 3, \dots, r,\ \bar\sigma_3\bar\sigma_2(\bar\sigma_3\bar\sigma_1)^{-1},\ \bar\sigma_1\bar\sigma_3\bar\sigma_2(\bar\sigma_1\bar\sigma_3\bar\sigma_1)^{-1},\ \bar\sigma_1\bar\sigma_3\bar\sigma_j(\bar\sigma_1)^{-1},\\ &j = 3, \dots, r,\ \bar\sigma_3\bar\sigma_1\bar\sigma_i(\bar\sigma_3)^{-1},\ i = 1,2,\ \bar\sigma_3\bar\sigma_1\bar\sigma_j(\bar\sigma_1\bar\sigma_3\bar\sigma_1)^{-1},\ j = 3, \dots, r,\\ &\bar\sigma_1\bar\sigma_3\bar\sigma_1\bar\sigma_i(\bar\sigma_1\bar\sigma_3)^{-1},\ i = 1,2, \text{ and } \bar\sigma_1\bar\sigma_3\bar\sigma_1\bar\sigma_j(\bar\sigma_3\bar\sigma_1)^{-1},\ j = 3, \dots, r\,\}\end{aligned}$$

generates $\widehat{H(1)}$.

This all simplifies once we go to the quotient by the relations defining $N$ in Lemma 3.1. For simplicity we do just the case $r = 4$. It is easy to see that the following subsets of $M$ and $\hat M$ respectively, generate the same quotient groups modulo $N$ as do $M$ and $\hat M$:

$$M' = \{\bar\sigma_1\bar\sigma_2,\ \bar\sigma_3\bar\sigma_1\bar\sigma_3,\ \bar\sigma_1\bar\sigma_3\bar\sigma_1,\ \bar\sigma_3\bar\sigma_2\bar\sigma_3,\ \bar\sigma_1\bar\sigma_4\bar\sigma_1\}; \text{ and}$$

$$\begin{aligned}\hat M' = \{&\bar\sigma_1\bar\sigma_2,\ \bar\sigma_1\bar\sigma_4\bar\sigma_3\bar\sigma_1, \bar\sigma_3\bar\sigma_1\bar\sigma_2\bar\sigma_3,\ \bar\sigma_3\bar\sigma_1\bar\sigma_4\bar\sigma_1\bar\sigma_3\bar\sigma_1,\\ &\bar\sigma_1\bar\sigma_3\bar\sigma_1\bar\sigma_3\bar\sigma_1\bar\sigma_3,\ \bar\sigma_1\bar\sigma_3\bar\sigma_1\bar\sigma_4\bar\sigma_1\bar\sigma_3\}.\end{aligned}$$

Indeed , only the 1st and 5th elements of $\hat M'$ are really needed. For example

$$\bar\sigma_1\bar\sigma_4\bar\sigma_3\bar\sigma_1 = \bar\sigma_1\bar\sigma_1\bar\sigma_2\bar\sigma_1 = \bar\sigma_2\bar\sigma_1 \bmod N.$$

This is the 1st element's inverse. And the inverse of the 5th times the 6th is

$$\bar\sigma_3\bar\sigma_1\bar\sigma_3\bar\sigma_4\bar\sigma_1\bar\sigma_3 = \bar\sigma_3\bar\sigma_1\bar\sigma_2\bar\sigma_1\bar\sigma_1\bar\sigma_3 = \bar\sigma_3\bar\sigma_4 \bmod N.$$

Thus we may regard $\pi_1(\hat X)$ as the quotient mod$N$ of the group generated by

$$\hat M'' = \{\bar\sigma_1\bar\sigma_2 \quad \text{and} \quad \bar\sigma_1\bar\sigma_3\bar\sigma_1\bar\sigma_3\bar\sigma_1\bar\sigma_3\}.$$

is reassuring to see that these two commute :

$$(\bar\sigma_1\bar\sigma_2)\bar\sigma_1\bar\sigma_3\bar\sigma_1\bar\sigma_3\bar\sigma_1\bar\sigma_3 = \bar\sigma_1\bar\sigma_3\bar\sigma_4\bar\sigma_3\bar\sigma_1\bar\sigma_3\bar\sigma_1\bar\sigma_3$$

$\bar\sigma_1\bar\sigma_3\bar\sigma_1\bar\sigma_2\bar\sigma_1\bar\sigma_3\bar\sigma_1\bar\sigma_3 \bmod N$, etc. Thus everything is in agreement with the computation that $\hat{X}) = 1$. Now consider $M'$.

The group generated by $M' \bmod N$ is the same as the group generated by $M'' =$ $_1\bar\sigma_2, \bar\sigma_1\bar\sigma_3\bar\sigma_1, \bar\sigma_3\bar\sigma_1\bar\sigma_3\}$. Denote this group by $H(1)'$. The group generated by it is clearly non-vial $\bmod N$, and since $g(X) = 0$, the group $H(1)'$ cannot possibly be $\pi_1(X)$ . Below, however, see that if we form the quotient by the torsion elements in $H(1)'$ we get $\pi_1(X)$. □

**he non-Galois case:** Now return to the group $H(1)$, a possibly nonnormal subgroup of $F_r$ derived from the Schreier construction. It is easy to interpret its quotient, $H(1)'$ , modulo the oup $N$ of Lemma 3.1 : Subgroups of $H(1)'$ are in one-one correspondence with covers $X' \to X$ th the property that the pullback over $\hat{X}$ (i.e., a connected component of the fiber product $\times_X X'$ ) is unramified over $X'$.

Since $g(X) = 0$ in Ex. 3.2, simple principle will help us to find the natural normal bgroup $N'$ of $H(1)'$ whose quotient will be trivial. The image of each subgroup $H$ of $H(1)'$ $H(1)'/N'$ should be the same as the image of $H \cap (\widehat{H(1)}/N)$ and $N'$ should be minimal with is property. In particular this applies to $H = H(1)'$. But in this case we use that fundamental oups of Riemann surfaces have no torsion. Therefore the fundamental group quotient will have torsion. Thus the image of the elements $\bar\sigma_1\bar\sigma_3\bar\sigma_1 = \alpha$ and $\bar\sigma_3\bar\sigma_1\bar\sigma_3 = \beta$, both of order dividing 2, ıst be 1. Just for the record, $\alpha\bar\sigma_1\bar\sigma_2 = \bar\sigma_1\bar\sigma_4\bar\sigma_1 = \bar\sigma_1\bar\sigma_2 \bmod N'$. Thus $\bar\sigma_1\bar\sigma_2$ has image $\bmod N'$ o of order dividing 2, and it too must be 1. We have now shown that $N' = H(1)'$ in Ex. 3.2. ıus we have established on general principles how to recognize that $\pi_1(X)$ is trivial.

For the general cover $X \to \mathbf{P}^1_x$ we show that the same idea works if each nontrivial cover $\to X$, fitting in a diagram $\hat{X} \to Y \to X$, is ramified. That is, replace $X$ by the maximal ramified cover $X^{\text{un}}$ of $X$ fitting between $\hat{X}$ and $X$.

**mma 3.3:** *The cover $X^{\text{un}}$ corresponds to the minimal subgroup $H^{\text{un}}$ of $G(1)$ with this property: e length of any orbit $O$ of $\sigma_i$ on the right cosets of $G(1)$ is the same as the lengths of each of e orbits of $\sigma_i$ on the right cosets of $H$ that comprise $O$, $i = 1, \ldots, r$. Finally, this condition is uivalent to the following :*

(3.3) *$\alpha \in G(1)$ if and only if $\alpha \in H$ as $\alpha$ runs over all elements of the form $g\sigma_i^k g^{-1}$ with $g \in G$, $k$ a divisor of $\text{ord}(\sigma_i)$, $i = 1, \ldots, r$.*

**oof:** Denote the cover of $\mathbf{P}^1_x$ given by any subgroup $H$ of $G(1)$ by $X_H \to \mathbf{P}^1_x$ so that $X_H \to$ $\to \mathbf{P}^1_x$ corresponds to the chain of subgroups $H \subseteq G(1) \subseteq G$. As in §1.2 let $\boldsymbol{x} = (x_1, \ldots, x_r)$ be branch points of the cover $\hat{X} \to \mathbf{P}^1_x$. From [Fr,2; p.147 expression (1.6)], a point $\boldsymbol{p}_H$ of $X_H$ ove $x_i$ corresponds to an orbit of $\sigma_i$ on the right cosets of $H$ in $G$, and the image $\boldsymbol{p}$ of $\boldsymbol{p}_H$ in $X$ responds to the extension of this orbit by $G(1)$ (i.e., by multiplication of the cosets for $H$ on left by $G(1)$).

Let $e(\boldsymbol{p}_H/\boldsymbol{p})$ (resp., $e(\boldsymbol{p}_H/x_i)$ and $e(\boldsymbol{p}/x_i)$) be the ramification index of $\boldsymbol{p}_H$ over $\boldsymbol{p}$ (resp., :.). Then, $e(\boldsymbol{p}_H/\boldsymbol{p}) \cdot e(\boldsymbol{p}/x_i) = e(\boldsymbol{p}_H/x_i)$ and this expresses that the orbit of $\sigma_i$ on the cosets of that corresponds to $\boldsymbol{p}_H$ has length $e(\boldsymbol{p}_H/\boldsymbol{p})$ times the length of the orbit of $\sigma_i$ on the cosets en by extension by $G(1)$. Of course, $X_H \to X$ is unramified if and only if $e(\boldsymbol{p}_H/\boldsymbol{p})=1$ for all ints $\boldsymbol{p}_H \in X_H$. Thus the condition in the lemma restates exactly that this cover is unramified. e restatement of the condition in (3.3) follows by noting these two points: $\alpha$ fixes an $H$-coset if and only if $g\alpha g^{-1} \in H$; and $\alpha$ has an orbit of length $k$ on $H$-cosets including $H$ if and only $k$ is the minimal integer $t$ such that $\alpha^t \in H$. It only remains to note that there is a minimal bgroup $H$ with this property. □

Again recall the natural surjective map $\delta : H(1)' \to G(1)$ induced from $\delta : F_r \to G(\boldsymbol{\sigma})$. henever there can be no confusion we denote the normal subgroup of $H(1)$ generated by torsion elements by **tor**. As a presented group it is generated by the image in $H(1)'$ of the set

$$T = \{\tau\alpha\tau^{-1} \mid \tau \in F_r,\ \alpha = \bar\sigma_i^k,\ k \text{ divides } \mathrm{ord}(\sigma_i),\ i = 1,...,r, \text{ and } \delta(\tau\alpha\tau^{-1}) \in G(1)\}.$$

Denote the subgroup $< \delta(\beta) \mid \beta \in T >$ of $G(1)$ by $G_{\mathbf{tor}}$. In Ex. 3.2 $G_{\mathbf{tor}} = G(1)$.

**Example 3.4:** $G_{\mathbf{tor}} \neq G(1)$. Let $G = S_n$, with $n$ even and at least 4, in its standa representation and take $\boldsymbol\sigma$ in the case where $\sigma_1 = (1\,2\ldots n)$, $\sigma_2 = \sigma_1^{-1}$, and all of the remain of the $\sigma$'s are 3-cycles which generate $A_n$. Then $G(1) = S_{n-1}$ (identified with the stabilizer 1), but $G_{\mathbf{tor}} = A_{n-1}$. Note, however, in the notation of Lemma 3.3 that $H^{\mathrm{un}} = A_{n-1}$. T is, suppose that $H$ is a subgroup of $G(1)$ containing $H^{\mathrm{un}}$ andJ $\sigma$ runs over 3-cycles in $G(1)$ t are conjugate to those in $\{\sigma_i,\ 1,\ldots,r\}$. By hypothesis these generate $A_{n-1}$. Then, according Lemma 3.3, $H\sigma = H$ for each of these. Thus $H$ contains all of these $\sigma$'s and so equals $A_{n-1}$.

The conclusion of Ex. 3.4 holds in general.

**Theorem 3.5:** *Use the previous notation with $\widehat{H(1)}$ the maximal normal subgroup of $F_r$ co tained in $H(1)$. We may compute $\pi_1(X)$ as the quotient $H(1)'/N'$ where $N'$ is the small normal subgroup of $H(1)'$ with the property that the induced map from $\widehat{H(1)}$ is surjective o $H(1)'/N'$. In particular the quotient $H(1)'/$**tor** of $H(1)'$ by **tor** maps surjectively to $\pi_1($ Furthermore, this is an isomorphism if and only if in the natural map $\delta : H(1)' \to G(1)$, image of **tor** is surjective. This holds if and only if $H^{\mathrm{un}} = G(1)$.*

**Proof:** Everything, except the conclusion that $H^{\mathrm{un}} = G(1)$ if and only if **tor** maps surjectiv onto $G(1)$ by $\delta$, follows from the previous discussion. But Lemma 3.3 shows that the ima under $\delta$ of the elements of $T$ generate $H^{\mathrm{un}}$, and thus $\delta(\mathbf{tor}) = H^{\mathrm{un}}$. □

## §3.2. $H_{\boldsymbol\sigma}$ ACTION ON $\pi_1(X)$

In the notation of §1.2 the subgroup $H_{\boldsymbol\sigma}$ is the stabilizer in $H(r)$ of an element $\boldsymbol\sigma \in \mathrm{Ni}(\mathbf{C})^{ab}_T$. To s the $H_{\boldsymbol\sigma}$ action explicitly recall from Theorem 3.5 that $\pi_1(X)$ is identified with $H(1)N/N = H($ modulo $N'$, where $N'$ is the minimal normal subgroup of $H(1)'$ such that the induced map fro $\widehat{H(1)}$ is surjective.

Denote the normal subgroup of $F_r$ generated by $\bar\sigma_1\cdots\bar\sigma_r$ by $N_0$. Then we identify $\pi_1(X - \phi^{-1}(\boldsymbol{x}))$ with $H(1)N_0/N_0$. Consider $Q \in H(r)$. It is in the group generated by $Q_1,\ldots,Q_r$ as in §1.2. Let $Q$ act on $\bar{\boldsymbol\sigma} = (\bar\sigma_1,\ldots,\bar\sigma_r)$ through the same formula as in (1.7). Clearly $Q$ ma $N_0$ into itself. Furthermore, consider the application of $Q$ to one of the generators (notation in §3.1) $rs\rho(rs)^{-1} \in M$ of $H(1)$. Assume now that $Q \in H_{\boldsymbol\sigma}$. For simplicity assume also th $(\boldsymbol\sigma)Q = \boldsymbol\sigma$. (Modification for the case when $\gamma(\boldsymbol\sigma)Q\gamma^{-1} = \boldsymbol\sigma$ for some $\gamma$ in the normalizer of $G$ easy.) Then the image of $(rs\rho(rs)^{-1})Q$ in $G(\boldsymbol\sigma)$ is the same as the image of $rs\rho(rs)^{-1}$. Since $H($ consists of those elements whose image is in $G(1)$, clearly $Q$ maps $H(1)$ into itself. Similarly, maps $\widehat{H(1)}$ into itself.

Let $N^*$ be the smallest normal subgroup of $H(1)$ containing $N$ (discussion prior to Lemm 3.3) such that $\widehat{H(1)}/N^* = H(1)/N^*$. We have induced an action of $Q$ on $\pi_1(X)$ if we show th $Q$ maps $N^*$ into itself. But $Q(N^*)$ clearly has the same properties as does $N^*$, once it has bee established that $Q(N) = N$. This reduces to showing that $\bar\sigma_i^{\mathrm{ord}(\sigma_i)}$ is in $N$, $i = 1,\ldots,r$. Goi back to the generating elements of $H(r)$, $(\bar\sigma_i)Q = \alpha\bar\sigma_j\alpha^{-1}$ for some $\alpha \in F_r$ and some $j$. Sin the image of $(\bar\sigma_i)Q$ and $\bar\sigma_i$ in $G(1)$ are the same, $\mathrm{ord}(\sigma_j) = \mathrm{ord}(\sigma_i)$. As $N$ is a normal subgrou of $F_r$, $(\bar\sigma_i^{\mathrm{ord}(\sigma_i)})Q = \alpha\bar\sigma_j\alpha^{-1}$ is in $N$. The following includes a summary.

**Theorem 3.6:** *Suppose that $T(\mathbf{C})_{\boldsymbol m} = X$ is one of the fibers of the family $\mathcal{F}(\mathbf{C})$, and that th fiber , as a cover of $\mathbf{P}^1_x$ , has $\boldsymbol\sigma$ as a description of its branch cycles. The subgroup $H_{\boldsymbol\sigma}$ of $H($ that leaves the image of $\boldsymbol\sigma$ in $\mathrm{Ni}(\mathbf{C})^{ab}_T$ fixed, induces an action on $\pi_1(X)$ through the action*

*$H(r)$ on $\bar{\sigma}$. This action can be identified with the usual (Picard-Lefshetz) monodomy action of the fundamental group of a parameter space on the fibers of a smooth complex analytic family. Furthermore, in the case that $g = 1$ or $2$, the image of $H_{\sigma}$ in $\mathrm{Aut}(H_1(X, \mathbf{Z}))$ is a finite group if and only if $\Psi(\mathcal{H}(\mathbf{C}), \mathcal{M}_g) : \mathcal{H}(\mathbf{C}) \to \mathcal{M}_g$ is constant.*

**Proof:** The action is described above. The identification with the usual monodromy action follows from [Fr,1; §4] which shows the effect of the $Q_i$'s on generating paths (e.g., Figure 1) of $\pi_1(\mathbf{P}^1_x - \boldsymbol{x})$. If these are represented by $\bar{\sigma}$ the action is given by (1.7). The induced action on paths representing $\pi_1(X - \phi_{-1}(\boldsymbol{x}))$ follows from the uniqueness up to homotopy of the natural fundamental group action.

The final statement comes by identifying generators of $H_{\sigma}$ with the Picard-Lefshetz transformation around a branch at $\infty$, in the language of [Gr; §6, especially Theorem 6.4—The removable singularity theorem]. If each of these generators is of finite order on $H_1(X, \mathbf{Z})$, then $\Psi(\mathcal{H}(\mathbf{C}), \mathcal{M}_g) : \mathcal{H}(\mathbf{C}) \to \mathcal{M}_g$ extends to $\overline{\mathcal{H}(\mathbf{C})} \to \mathcal{M}_g$ where $\overline{\mathcal{H}(\mathbf{C})}$ is a nonsingular projective compactification of $\mathcal{H}(\mathbf{C})$. But, in the case that $g = 1$ or $2$, $\mathcal{M}_g$ is an affine open subset of a projective variety $\overline{\mathcal{M}}_g$ [M; p.25]. By Chow's theorem the image of $\overline{\mathcal{H}(\mathbf{C})}$ is a projective subvariety of $\mathcal{M}_g$. Thus, unless it is just a point it must meet one of the divisors in $\overline{\mathcal{M}}_g \setminus \mathcal{M}_g$, contrary to our information. The converse of the last statement is much easier. □

**Remark:** For $g \geq 3$, $\mathcal{M}_g$ is not affine (e.g., it contains projective curves), but the "coalescing of branch points" argument (as in [FrG; §5.2] or Statement 2.16) often works to check for the possibility of extending $\mathcal{H}(\mathbf{C}) \to \mathcal{M}_g$ to a map into $\mathcal{M}_g$ along a specific branch at $\infty$ of $\mathcal{H}(\mathbf{C})$. □

## §3.3. ENDOMORPHISMS AND THE LEFSCHETZ TRACE FORMULA

Now consider the group $\mathrm{Aut}(X/\mathbf{P}^1_x)$ of automorphisms of the cover $\phi : X \to \mathbf{P}^1_x$. In terms of branch cycles, this is naturally identified with the centralizer, $\mathrm{Cen}_{S_n}(G(\boldsymbol{\sigma}))$, of $G(\boldsymbol{\sigma})$ in $S_n$ [Fr,2; Lemma 2.1]. When we take $X$ to be $\hat{X}$ (i.e., the cover is Galois) this is the regular representation of $G(\boldsymbol{\sigma})$ and $n = \hat{n} = |G(\boldsymbol{\sigma})|$ [Fr,2; Lemma 2.1]. This induces an action of $\mathrm{Aut}(\hat{X}/\mathbf{P}^1_x)$ on $H_1(\hat{X}, \mathbf{Z})$ (and $H_1(\hat{X}, \mathbf{Q})$) which is known to be faithful [FaK; p.253]. Thus the group ring $A = \mathbf{Q}[\mathrm{Aut}(\hat{X}/\mathbf{P}^1_x)]$ acts faithfully on $H_1(\hat{X}, \mathbf{Q})$.

For each $\alpha \in \mathrm{Aut}(\hat{X}/\mathbf{P}^1_x)$ choose a (homotopy class of) path $\bar{\sigma}_\alpha$ on $\hat{X}$ with initial point $p_1$ and end point $\alpha(p_1)$. (Note that such choices would have already been made in applying Schreier's construction to compute the fundamental group of $\hat{X}$ in terms of branch cycles.) The following uses a number of classical results, including the Lefschetz trace formula: The alternating sum of the traces of an automorphism of a Riemann surface on the integral homology spaces is the number of fixed points of the automorphism [FaK; p.265].

**Principle 3.7:** *In the notation above, $\alpha$ acts on $H_1(\hat{X}, \mathbf{Q})$ by conjugation by $\bar{\sigma}_\alpha$. Denote the number of disjoint cycles of $\sigma_i$ (in the regular representation of $G$) that $\alpha$ centralizes by $t_i$. Then,*

*(3.4) the trace of the action of $\alpha$ on $H_1(\hat{X}, \mathbf{Q})$ is $2 - \sum_{i=1}^{r} t_i$.*

*We may effectively do the following: identify $H_1(X, \mathbf{Q})$ with a subspace of $H_1(\hat{X}, \mathbf{Q})$; and given $\beta \in A$, decide if $\beta$ maps $H_1(X, \mathbf{Q})$ into itself and in this case deduce whether the action of $\beta$ is nontrivial. Denote the elements of $A \otimes \mathbf{Q}$ that leave $H_1(X, \mathbf{Q})$ stable by $A_X$. If $\mathrm{Ni}(\mathbf{C})$ has full moduli dimension $3g - 3$, then $A_X = \mathbf{Q}$.*

Denote the divisor classes (not necessarily positive) of degree $k$ on $X$ (resp., $\hat{X}$) by $\mathrm{Pic}^k(X)$ (resp., $\mathrm{Pic}^k(\hat{X})$). Consider an element

$$c = \sum_{\tau \in \mathrm{Aut}(\hat{X}/\mathbf{P}^1_x)} a_\tau \tau \in A.$$

Then c induces a map $c^* : \mathrm{Pic}^1(\hat{X}) \to \mathrm{Pic}^k(\hat{X})$ where $k = \sum_{\tau \in \mathrm{Aut}(\hat{X}/\mathbf{P}^1_x)} a_\tau$ by mapping $\hat{\boldsymbol{p}}$ $\hat{X} \to \sum_{\tau \in \mathrm{Aut}(\hat{X}/\mathbf{P}^1_x)} a_\tau \tau(\hat{\boldsymbol{p}})$. Let $\psi : \hat{X} \to X$ be the natural map. Identify a point $\boldsymbol{p} \in X$ wit $\sum_{\tau \in G(1)} \tau(\hat{\boldsymbol{p}})$ by choosing any point $\hat{\boldsymbol{p}} \in \hat{X}$ lying above $\boldsymbol{p}$. Then we recognize $\mathrm{Pic}^1(X)$ as th image of $\mathrm{Pic}^1(\hat{X})$ under the map $c(\psi)^* : \mathrm{Pic}^1(\hat{X}) \to \mathrm{Pic}^k(\hat{X})$ with $c(\psi) = \sum_{\tau \in G(1)} \tau$. Suppo $c \in A$ satisfies $c \cdot c(\psi) = c(\psi) \cdot c$. Then there exists a commutative diagram

$$\begin{array}{ccc} \mathrm{Pic}^1(\hat{X}) & \xrightarrow{c(\hat{\psi})} & \mathrm{Pic}^1(X) \\ \downarrow{\scriptstyle c} & & \downarrow{\scriptstyle c_0} \\ \mathrm{Pic}^k(\hat{X}) & \xrightarrow{c(\psi)} & \mathrm{Pic}^k(X) \end{array} \tag{3.5}$$

where $c_0$ is defined by application to $c(\psi)(\hat{\boldsymbol{p}})$ where it yields $c(\psi)(c(\hat{\boldsymbol{p}}))$. Note that $c_0$ induces a endomorphism on $\mathrm{Pic}^0(X)$.

At any time, up to isogeny, this allows us to identify $\mathrm{Pic}^0(X)$ with the image of $c(\hat{\psi})$ : $\mathrm{Pic}^0(\hat{X})$. In practical examples (e.g., [FrG; §5.3]) we check if there exists a $c$ such that $c_0$ induc a nontrivial endomorphism (i.e., not in $\mathbf{Z}$) of $\mathrm{Pic}^0(X)$) by using the Lefschetz trace formu (Principal 3.7). That is, we decompose $H_1(\hat{X}, \mathbf{Q})$ in terms of representations of $\mathrm{Aut}(\hat{X}/\mathbf{P}^1_x)$ ar then we apply $c(\psi)$ to these modules. If any of the resulting modules have an induced nontrivi endomorphism, then the Nielsen class is not of full moduli dimension (c.f. Statement 2.8).

**Problem 3.8:** *Apply the procedure above to the case when the Nielsen class* $\mathrm{Ni}(\mathbf{C})^{ab}_T$ *is give by* $G = A_n$ *and* $\mathbf{C}$ *has all conjugacy classes equal to that of a 3-cycle. What are the values of for which this example has full moduli dimension?*

This particular example appears quite generally in Thompson's program (Statement 2.6

## Bibliography


[ArC] E. Arbarello and M. Cornalba, The number of $g^1_d$'s on a general $d$-gonal curve, and th unirationality of the Hurwitz spaces of 4-gonal and 5-gonal curves, *Math. Ann.* **25** (1981), 341–362.

[AM] M. Artin and B. Mazur, Comments by Artin and Mazur in the introduction to [Z]

[B] R. Bourgne at J.-P. Azra, Écrits at Mémoires Mathématiques D' Évariste Galois, *Gau hiers–Villars et Cie*, (1962), Paris.

[FaK] H. Farkas and I. Kra, Riemann Surfaces, *Graduate texts in Math* **71** (1980), Springe Verlag, New York.

[Fr,1] M. Fried, Fields of definition of function fields and Hurwitz families ..., *Comm. in Al* **5(1)** (1977), 17–82.

[Fr,2] M. Fried, Galois group and complex multiplication, *TAMS* **235** (1978), 141–163.

[Fr,3] M. Fried, Poncelet correspondences; Finite correspondences; Ritt's Theorem; and th Griffiths–Harris configuration for quadrics, *J. of Alg.* **54** (1978), 467–493.

[FrG] M. Fried and R. Guralnick, On uniformization of generic curves of genus $g(< 6)$ b radicals, preprint.

[FrJ] M. Fried and M. Jarden, Field Arithmetic, *Springer, Ergebnisse,* **11** (1986).

[Gr] P. Griffiths, Periods of integrals on algebraic manifolds;...,*BAMS* **76** (1970), 228–296.

[Gro] A. Grothendieck, Géométrie formelle et géométrie algébrique, *Seminaire Bourbaki* **t.1** **182** (1958/59).

[GTh] R. Guralnick and J.G. Thompson, On genus zero covers, in preparation.

[KL] S. Kleiman and D. Laksov, Another proof of the existence of special divisors,*Acta Mat* **132** (1974), 163–175.

[L] S. Lefschetz, A Theorem on Correspondences on Algebraic Curves, *Amer. J. Math.* **5** (1928), 159–166.

[M] D. Mumford, Curves and their Jacobians, *University of Michigan Press* (1976)

[R] J. Ries, The Prym variety for a cyclic unramified cover of a hyperelliptic surface, *J. für die reine und angew.* **340** (1983), 59- 69.

[Ri] J.F. Ritt, On algebraic functions which can be expressed in terms of radicals, *TAMS* 24 (1922), 21–30.

[Z] O. Zariski, Collected Papers Volume III. Topology of Curves and Surfaces, and Special Topics in the Theory of Algebraic Varieties, *MIT Press* (1978).

Mike Fried

Department of Mathematics
201 Walker Hall
University of Florida
Gainesville, Fl 32611

Contemporary Mathematics
Volume **89**, 1989

# Labeled trees and the algebra of differential operators

Robert Grossman*
University of California, Berkeley

Richard G. Larson†
University of Illinois at Chicago

This paper is concerned with the effective symbolic computation of operators under composition. Examples include differential operators under composition and vector fields under the Lie bracket. A basic fact about such operators is that, in general, they do not commute. However, they are often rewritten in terms of other operators which do commute. If the original expression enjoys a certain symmetry, then naive rewriting requires the computation of terms which in the end cancel.

In this paper we analyse data structures consisting of formal linear combinations of rooted labeled trees. We define a multiplication on rooted labeled trees, thereby making the vector space spanned by the set of these data structures into an associative algebra. We then define algebra homomorphisms from the original algebra of noncommuting operators to this algebra of trees, and from this algebra of trees to the algebra of differential operators on $R$. The cancellation which occurs when noncommuting operators are expressed in terms of commuting ones occurs naturally when the operators are represented using this data structure. This leads to an algorithm which, for operators which are derivations, speeds up the computation exponentially in the degree of the operator.

First consider a concrete example: Fix three vector fields $E_1, E_2, E_3$ in

---

*The first author is a National Science Foundation Postdoctoral Research Fellow.
†This paper was written while the second author was on sabbatical leave at the University of California, Berkeley.


0271-4132/89 $1.00 + $.25 per page

$\mathbf{R}^N$ with polynomial coefficients $a_i^j$:

$$E_i = \sum_{j=1}^{N} a_i^j \frac{\partial}{\partial x_j}, \qquad \text{for } i = 1, 2, 3.$$

Considering these vector fields as first-order differential operators, it is natural to form higher-order differential operators from them, such as the third-order differential operator

$$p = E_3E_2E_1 - E_3E_1E_2 - E_2E_1E_3 + E_1E_2E_3.$$

Writing this differential operator in terms of $\partial/\partial x_1, \ldots, \partial/\partial x_N$ yields a first-order differential operator because the symmetry of the expression $p$ causes all higher order terms to cancel.

In this paper we analyse an algorithm for computing such differential operators which does not involve the explicit computation of the higher order terms which cancel. In the example above, the naive computation would require the computation of $24N^3$ terms, while the algorithm we describe here would involve just the computation of the $6N^3$ terms which do not cancel.

We conclude this introduction with some remarks, and a theorem.

1. In actual applications, expressions possessing symmetry arise more often than not. For example, Lie brackets of vector fields possess a great deal of symmetry, as does the Laplacian $E_1E_1 + E_2E_2 + E_3E_3$ built from the vector fields. The algorithm we discuss is designed to take advantage of such symmetries, if they are present, without the necessity of explicitly identifying the symmetries.

2. Once a set of data structures has been given an algebraic structure, it becomes natural to view a simplification algorithm as simply the factorization of a map from the original algebra of interest through a map to the algebra of these data structures. This is the idea which is the basis of the algorithm we describe. We expect that this idea will find application elsewhere.

3. The space of operators on a linear space is not only an algebra but also a coalgebra; that is, it is the dual of an algebra. The algebra of data structures mentioned above also has a coalgebra structure. Although this fact plays a relatively minor role in the simple algorithms discussed

in this paper, it does play a crucial role for other algorithms we have studied.

4. The algorithm described here lends itself to parallelization. We are currently investigating this.

5. See [2] and [3] for previous work on the simplification of expressions. The collection [1] contains several articles and many references pertaining to symbolic computation; in particular, the use of derivations in symbolic computations occurs in integration algorithms. For unexplained notions concerning data structures, see [6].

Let $\mathrm{Cost}_{\mathrm{A}}(p)$ denote the cost of using algorithm A to simplify $p$. In the final section we show that there is an algorithm using trees whose work compares to the amount of work required by naive substitution as follows.

**Theorem 1** *Assume*

1. *$p$ is the sum of $2^{m-1}$ terms, each homogenous of degree $m$;*

2. *the differential operator obtained from the substitution (1) below is a linear differential operator of degree 1;*

3. *$m, N \to \infty$ in such a way that $2^{m-2}m < N^m$.*

*Then*

$$\frac{\mathrm{Cost}_{\mathrm{TREES}}(p)}{\mathrm{Cost}_{\mathrm{NAIVE}}(p)} = O\left(\frac{1}{m2^{m-1}}\right)$$

Observe that a Lie bracket of length $m$ on $\mathbf{R}^N$ with $2^{m-2}m < N^m$ satisfies the hypotheses of this theorem.

# 1 Derivations, trees, and algebras

In this section we collect some facts about derivations, trees, and the algebras they generate. Let $R$ be a commutative algebra with unit over the field $k$. (Throughout this paper $k$ is a field of characteristic 0.) A *derivation* of the algebra $R$ is a linear map $D$ of $R$ to itself satisfying

$$D(ab) = aD(b) + bD(a), \qquad \text{for all } a, b \in R.$$

Let $D_1, \ldots, D_N$ be $N$ commuting derivations of $R$, that is, for $i$, $j = 1, \ldots, N$,

$$D_i D_j a = D_j D_i a, \qquad \text{for all } a \in R.$$

Suppose that we are also given $M$ derivations $E_1, \ldots, E_M$ of $R$ which can be expressed as $R$-linear combinations of the derivations $D_i$; that is, for $j = 1, \ldots, M$,

$$E_j = \sum_{\mu=1}^{N} a_j^{\mu} D_{\mu}, \qquad \text{where } a_j^{\mu} \in R. \tag{1}$$

We are interested in writing higher-order derivations generated by the $E_1, \ldots, E_M$ in terms of the commuting derivations $D_1, \ldots, D_N$. More formally, let $k{<}E_1,\ldots,E_M{>}$ denote the free associative algebra in the symbols $E_1, \ldots, E_M$ and let $\mathbf{Diff}(D_1, \ldots, D_N; R)$ denote the space of linear differential operators generated by the $D_\mu$ with coefficients from $R$; that is, $\mathbf{Diff}(D_1, \ldots, D_N; R)$ consists of all finite formal expressions

$$L = \sum_{\mu_1=1}^{N} a_{\mu_1} D_{\mu_1} + \sum_{\mu_1,\mu_2=1}^{N} a_{\mu_1\mu_2} D_{\mu_2} D_{\mu_1} + \cdots$$

where $a_{\mu_1}, a_{\mu_1\mu_2}, \ldots \in R$. We let

$$\chi : k{<}E_1,\ldots,E_M{>} \longrightarrow \mathbf{Diff}(D_1, \ldots, D_N; R)$$

denote the map which sends $p \in k{<}E_1,\ldots,E_M{>}$ to the linear differential operator $L = \chi(p)$ obtained by performing the substitution (1), and simplifying using the fact that the $D_\mu$ are derivations of $R$.

The Lie bracket of two derivations $E_1$ and $E_2$ of $R$ is defined to be

$$[E_1, E_2] = E_1E_2 - E_2E_1.$$

It is a basic fact that $[E_1, E_2]$ is also a derivation. We define its length to be 2, and the length of the $E_i$ to be 1. In general the *length* of a Lie bracket is defined to be $k + l$ if it can be written $[F_1, F_2]$, where $F_1$ is a Lie bracket of length $k$ and $F_2$ is a Lie bracket of length of $l$.

We conclude this section by describing an algebra associated with the family of trees. By a tree we mean a rooted finite tree [6]. If $\{E_1, \ldots, E_M\}$ is a set of symbols, we will say a tree is *labeled with* $\{E_1, \ldots, E_M\}$ if every node of the tree other than the root has an element of $\{E_1, \ldots, E_M\}$ assigned to it. We denote the set of all trees labeled with $\{E_1, \ldots, E_M\}$ by $LT(E_1, \ldots, E_M)$, and the subset consisting of rooted trees containing $m+1$ nodes by $LT_m(E_1, \ldots, E_M)$. Let $k\{LT(E_1, \ldots, E_M)\}$ denote the vector space over $k$ with basis $LT(E_1, \ldots, E_M)$. We show now that this vector space is an algebra.

We define the multiplication in $k\{LT(E_1, \ldots, E_M)\}$ as follows. Since the set of labeled trees is a basis for this vector space, it is sufficient to describe the product of two labeled trees. Suppose $t_1$ and $t_2$ are two labeled trees. Let $s_1, \ldots, s_r$ be the children of the root of $t_1$. If $t_2$ has $n+1$ nodes (counting the root), there are $(n+1)^r$ ways to attach the $r$ subtrees of $t_1$ which have $s_1, \ldots, s_r$ as roots to the labeled tree $t_2$ by making each $s_i$ the child of some node of $t_2$, keeping the original labels. The product $t_1 t_2$ is defined to be the sum of these $(n+1)^r$ labeled trees. It can be shown that this product is associative, and that the tree consisting only of the root is a multiplicative identity. This implies

**Theorem 2** *The vector space $k\{LT(E_1, \ldots, E_M)\}$ is an associative algebra over $k$.*

In fact much more is true: there is a comultiplication and counit so that $k\{LT(E_1, \ldots, E_M)\}$ becomes a cocommutative graded conected Hopf algebra [5].

Let $k{<}E_1, \ldots, E_M{>}$ denote the free associative algebra on the set of symbols $\{E_1, \ldots, E_M\}$. Then there is an algebra homomorphism

$$\phi : k{<}E_1, \ldots, E_M{>} \longrightarrow k\{LT(E_1, \ldots, E_M)\}.$$

The map $\phi$ sends $E_i$ to the labeled tree with two nodes: the root, and a child of the root labeled with $E_i$; it is then extended to all of $k{<}E_1, \ldots, E_M{>}$ using the fact that it is an algebra homomorphism.

# 2 Simplification of higher order derivations

In this section we define a map

$$\psi : k\{LT(E_1, \ldots, E_M)\} \longrightarrow \mathbf{Diff}(D_1, \ldots, D_N; R).$$

We do this follows: Given a labeled tree $t \in LT_m(E_1, \ldots, E_M)$, assign the root the number 0, and assign the remaining nodes the numbers $1, \ldots, m$. Henceforward we identify the node with the number assigned to it. To each node $k > 0$ we associate a summation index $\mu_k$. Write $\mu = (\mu_1, \ldots, \mu_m)$. For $t \in LT_m(E_1, \ldots, E_M)$, let $k$ be a node of $t$, labeled with $E_{\gamma_k}$ if $k > 0$, and suppose that $l, \ldots, l'$ are the children of $k$. Let

$$R(k; \mu) = \begin{cases} D_{\mu_l} \cdots D_{\mu_{l'}} a_{\gamma_k}^{\mu_k} & \text{if } k \text{ is not the root;} \\ D_{\mu_l} \cdots D_{\mu_{l'}} & \text{if } k \text{ is the root.} \end{cases}$$

Note that if $k > 0$, then $R(k;\mu) \in R$. For $t \in LT_m(E_1, \ldots, E_M)$, define

$$\psi(t) = \sum_{\mu_1,\ldots,\mu_m=1}^{N} R(m;\mu)\cdots R(1;\mu)R(0;\mu).$$

Extend $\psi$ to all of $k\{LT(E_1,\ldots,E_M)\}$ by linearity.

The following proposition describes a fundamental property of the map $\psi$. Note that this proposition is an example of simplification by factoring $\chi$ through the set of labeled trees: we will see that often $\psi$ and $\phi$ together are cheaper to compute than $\chi$.

**Proposition 3**

$$\chi = \psi \circ \phi.$$

The proof is a straightforward verification and is contained in [4].

We now consider the cost of writing an expression composed of noncommuting operators in terms of commuting operators. We make the following asssumptions: $p \in k\langle E_1,\ldots,E_M\rangle$ is a sum of $l$ terms of degree $m$; the cost of a multiplication is one unit and the cost of a differentiation is one unit; the cost of an addition is zero units; and the cost of adding a node to a tree is one unit, so that the cost of building a tree $t \in LT_m(E_1, \ldots, E_M)$ is $m$ units. The costs given in the following three propositions ignore lower-order terms; their proofs are straightforward.

**Proposition 4** *$\chi(p)$ contains $l\,m!\,N^m$ terms, and the cost of computing $\chi(p)$ is $2lm\,m!\,N^m$.*

**Proposition 5** *The cost of computing $\sum_{i=1}^{l}\phi(p_i)$ is $lm\,m!$.*

**Proposition 6** *Let $\sigma = \phi(p)$, and denote by $|\sigma|$ the number of labeled trees with non-zero coefficients in $\sigma$. Then the cost of computing $\psi(\sigma)$ is $2m|\sigma|N^m$.*

Combining these three propositions gives

**Theorem 7** *Under the assumptions above, $\mathrm{Cost}_{\mathrm{NAIVE}}(p)$, the cost of computing $\chi(p) = \sum_{i=1}^{l}\chi(p_i)$ using naive substitution, is*

$$2lm\,m!\,N^m.$$

*On the other hand, $\mathrm{Cost}_{\mathrm{TREES}}(p)$, the cost of computing $\psi\circ\phi(p)$, is*

$$lm\,m! + 2m|\sigma|N^m.$$

Theorem 1 now follows.

# References


[1] B. Buchberger et. al. , "Computer algebra: symbolic and algebraic computation," Springer, Wien, 1983.

[2] B. Buchberger and R. Loos, *Algebraic Simplification,* in "Computer algebra: symbolic and algebraic computation," ed. B. Buchberger et. al., Springer, Wien, 1983, 11 - 43.

[3] B. F. Caviness, *On canonical forms and simplification,* J. Assoc. Computing Machinery **17** (1970), 385-396.

[4] R. Grossman, *Evaluation of expressions involving higher order derivations,* Center For Pure and Applied Mathematics, PAM - 367, University of California, Berkeley, submitted for publication.

[5] R. Grossman and R. Larson, *Hopf-algebraic structures of families of trees,* J. Algebra, to appear.

[6] R. E. Tarjan, "Data Structures and Network Algorithms," SIAM, Philadelphia, 1983.

Contemporary Mathematics
Volume **89**, 1989

# COMPUTING EDGE-TOUGHNESS AND FRACTIONAL ARBORICITY

by

Arthur M. Hobbs
Department of Mathematics
Oakland University
Rochester, Michigan 48309*

*ABSTRACT.* The relationships between two functions associated with the edge-connectivity of a graph are presently under investigation. One, introduced by Gusfield [17] in reciprocal form and called here the "edge-toughness" $\eta$, measures the global strength of a graph; the other, introduced by Catlin, Hobbs, and Lai [6] and called the "fractional arboricity " $\gamma$, measures a graph's local strength. The present paper has two purposes. One is to summarize the discoveries made so far with regard to these functions and to show how they are useful in network design. The second is to provide an algorithm for computing these functions for a matroid which is polynomial-time in tests of independence. This algorithm is then specialized to graphs.

The terminology in this paper comes from [2] and [33]. As defined in Bondy and Murty [2], $\omega(G)$ is the number of components of the graph $G$ , and $\kappa'(G)$ is the edge-connectivity of $G$. Unless otherwise stated, the graphs in this paper are connected.

For a nontrivial graph $G$, define the <u>edge</u> – <u>toughness</u> $\eta(G)$ by

$$\eta(G) = \min_{S \subseteq E(G)} \frac{|S|}{\omega(G-S) - \omega(G)} \tag{1}$$

where the minimum is over all subsets $S \subseteq E(G)$ such that $\omega(G-S) > \omega(G)$. This function was first introduced by Gusfield [17] in reciprocal form, and was later studied in a more general form by Cunningham [10].

*on leave from Texas A & M University, College Station, Texas 77843


0271-4132/89 $1.00 + $.25 per page

In 1961, C. St. J. A. Nash-Williams [24] and W. T. Tutte [31] produced a formula for the size of the largest set of edge-disjoint spanning trees possible in a given graph. Another proof was given in 1976 by Bollobás [1]. In terms of $\eta(G)$, this result is:

*THEOREM 1* [24], [31], [1]: In any nontrivial connected graph $G$, the maximum number of edge-disjoint spanning trees is $\lfloor \eta(G) \rfloor$.

For a nontrivial graph $G$, define

$$\gamma(G) = \max_{H \subseteq G} \frac{|E(H)|}{|V(H)| - \omega(H)} \tag{2}$$

where the maximum runs over all subgraphs $H$ with $|V(H)| > \omega(H)$. This function is suggested by formulas in papers of Tomizawa [30] and Narayanan and Vartak [22], [23], but it was first explicitly defined in [6].

Nash-Williams, in a 1964 paper [25], noticed that the 1961 work could be modified to produce what is called the " arboricity" of a graph. In terms of $\gamma(G)$ he showed:

*THEOREM 2* [25]: The minimum number of forests whose union contains the edge-set of a nontrivial graph $G$ is $\lceil \gamma(G) \rceil$.

The number $\lceil \gamma(G) \rceil$ is called the arboricity of $G$.

There is now literature in matroid theory ([3], [10], [11], [12], [16], [19], [22], [23], [30], [33]), directed graphs ([10], [14], [15], [29],) hypergraphs ([28]) and infinite graphs ([27]) extending the results of Tutte and Nash-Williams.

That "edge-toughness" is an appropriate term for $\eta(G)$ is seen first by considering the meaning of the number for a given graph. If we are interested in splitting a graph $G$ into $k + \omega(G)$ components by the removal of edges, the definition of $\eta(G)$ tells us that we must remove at least $k\eta(G)$ edges to achieve this result. Thus $\eta(G)$ measures how tough it is to break up $G$. Gusfield [17] introduced the parameter $T = 1/\eta(G)$. In connection with network design, his comment on the function is worth quoting: "It seems reasonable that $T$ can be used as a measure of vulnerability; the larger the ratio [i.e., the smaller the value of $\eta$], the more vulnerable is the graph to large amounts of disconnection for few edge deletions." Cunningham [10] generalized Gusfield's function to the strength $\sigma(G,s)$ of graph $G$, where $s$ is a function assigning a positive weight to each edge of $G$. Cunningham also generalized this function to matroids and provided a polynomial-time algorithm for computing its value.

Catlin has recently shown [4] that edge-toughness allows a deeper characterization of the edge-connectivity $\kappa'(G)$:

*THEOREM 3* [4]: Let $G$ be a graph, let $k \geq |E(G)|$ be an integer, and let $\Xi_k$ be the collection of all k-element subsets of $E(G)$. Then

(a) $\kappa'(G) \geq 2k$ if and only if for all $E' \in \Xi_k$, $\eta(G - E') \geq k$; and

(b) $\kappa'(G) \geq 2k + 1$ if and only if for all $E' \in \Xi_k, \eta(G - E') > k$.

It is well-known that $\kappa'(G)$ is characterized by Menger's Theorem ([2], p.204). For $E' \subseteq E(G)$, the maximum number of edge–disjoint spanning trees of $G - E$ is $\lfloor \eta(G - E') \rfloor$ by Theorem 1. Hence, as observed by Catlin [4], the first part of Theorem 3 can be restated in terms of edge-disjoint spanning trees of $G - E'$ as:

*COROLLARY 3A* [4]: $\kappa'(G) \geq 2k$ if and only if for all $E' \in \Xi_k$, the number of edge-disjoint spanning trees of $G - E'$ is at least $k$.

This restatement strengthens the following result of Kundu [20] (rediscovered by Gusfield [17]):

*COROLLARY 3B* [20], [17]: If a graph $G$ is $2k$–edge–connected, then $G$ has $k$ edge-disjoint spanning trees.

The following two theorems improve Theorems 1 and 2, which are the case $t = 1$ of these theorems. Theorem 4 was first proved implicitly by Cunningham ([10], page 551), and Theorem 5 is a generalization of a theorem of Narayanan and Vartak ([22], Theorem 8).

*THEOREM 4* [10], [6]: For any nontrivial graph $G$ and any natural numbers $s$ and $t$ we have $\eta(G) \geq s/t$ if and only if $G$ has a family $F$ of $s$ spanning trees such that each

each edge of $G$ lies in at most $t$ trees of $F$.

*THEOREM 5* [6], [22]: For any nontrivial graph $G$ and any natural numbers $s$ and $t$, $\gamma(G) \leq s/t$ if and only if there is a family $F$ of $s$ spanning trees in $G$ such that each edge of $G$ lies in at least $t$ trees of $F$.

As a result of Theorem 5, we may reasonably call $\gamma(G)$ the fractional arboricity of $G$.

Let $H$ be a subgraph of $G$. We define $G/H$ to be the graph obtained from $G$ by contracting each of the edges of $H$ to a vertex and deleting any resulting loops. Then we have:

*LEMMA 6* [6]: If $G$ has a subgraph $H$ such that $\eta(H) > \eta(G)$, then $\eta(G) = \eta(G/H)$.

This and a simple computation leads to the helpful computational corollary:

*COROLLARY 6A*: Let $G$ be a connected graph with blocks $B_1, B_2, \ldots, B_c$. Then $\eta(G) = \min_i \eta(B_i)$ and $\gamma(G) = \max_i \gamma(B_i)$.

The duality of $\eta$ and $\gamma$ is indicated by the following theorem. A more direct but less useful duality is shown by Theorem 15 in the section on matroids.

*THEOREM 7*: If $G$ is a graph of order $n > 1$, then

$$\text{(a) [6]} \quad \eta(G) \leq \frac{|E(G)|}{n-1} \leq \gamma(G);$$

$$\text{(b) [4]} \quad \gamma(G) = \max_{K_1 \neq H \subseteq G} \eta(H);$$

$$\text{(c) [4]} \quad \eta(G) = \min \gamma(H),$$

where the minimum in (c) runs over all nontrivial contractions $H$ of $G$.

Define a graph $G$ to be $\eta$-reduced if $G$ contains no subgraph $H$ with $\eta(H) > \eta(G)$. We note that if $G$ is not $\eta$-reduced, it is possible to pass through a sequence of contractions of subgraphs $H$ with $\eta(H) > \eta(G)$ to an $\eta$-reduced graph $G_0$; by Lemma 6, $\eta(G) = \eta(G_0)$.

We call $G_0$ an $\eta$-reduced graph of $G$. Narayanan and Vartak [22] use the term "molecular graph" for an $\eta$-reduced graph.

We define a $t$-covering of a graph $G$ to be a family $F$ of spanning trees of $G$ such that each $e \in E(G)$ is in at least $t$ members of $F$. We define a $t$-packing of $G$ to be a family $F$ of spanning trees of $G$ such that each edge $e \in E(G)$ is in at most $t$ members of $F$.

*THEOREM 8* [6]: For any graph G of order $n > 1$, the following are equivalent:

(a) $|E(G)| = \eta(G)(n-1)$;

(b) $|E(G)| = \gamma(G)(n-1)$;

(c) $\eta(G) = \gamma(G)$;

(d) $G$ is $\eta$-reduced;

(e) There is a function $f : \{2, 3, \ldots, n\} \to \mathbf{R}$, where $\mathbf{R}$ is the set of real numbers, such that

(1) $\dfrac{f(r)}{r-1} \leq \dfrac{f(n)}{n-1}$ for $2 \leq r \leq n$;

(2) $|E(G)| = f(n)$; and

(3) For any nontrivial subgraph $H$ of $G$, $|E(H)| \leq f(|V(H)|)$; and

(f) there is a natural number $t$ and a family $F$ of spanning trees of $G$ such that $F$ is both a $t$-covering and a $t$-packing.

Parts (d) and (f) of Theorem 8 were previously shown equivalent by Narayanan and Vartak [22]. Part (e) of Theorem 8 is particularly useful for proving part (a) of the next theorem.

*THEOREM 9* : The set of graphs $G$ satisfying $\eta(G) = \gamma(G)$ includes

(a) [6] plane triangulations;

(b) [6] for any natural number $k$, the edge-disjoint union of $k$ nontrivial spanning trees on a given set of vertices; and

(c) [22] edge-transitive graphs.

From these theorems, we have also a useful lower bound for $|E(G)|$ based on subgraphs of $G$. (This result was suggested by a problem of Paul Erdös [13].)

*THEOREM 10* [6]: Let $G$ be a nontrivial graph, let $T$ be a spanning tree of $G$ and let $F$ be a family of graphs. If each edge of $T$ is contained in a subgraph isomorphic to a member of $F$, then $\eta(G) \geq \inf_{H \in F} \eta(H)$.

*COROLLARY 10A* [6]: Let $G$ be a graph of order $n > 1$, let $T$ be a spanning tree of $G$, and let $F$ be a family of graphs. If each edge of $T$ is contained in a subgraph isomorphic to a member of $F$, then

$$|E(G)| \geq (\inf_{H \in F} \eta(H))(n-1).$$

For connected graphs, the functions $\eta$ and $\gamma$ take on rational values satisfying $\gamma(G) \geq \eta(G) \geq 1$. Are all possible combinations of rational numbers achieved by graphs? The answer in the affirmative was provided by Catlin, Foster, Grossman, and Hobbs [7] by constructing graphs satisfying the following theorem:

*THEOREM 11* [7]: If $x \geq y \geq 1$ are rational numbers, then there is a graph $G$ with $\gamma(G) = x$ and $\eta(G) = y$.

Unfortunately, the construction proving this theorem does not guarantee that $G$ will be simple if $x > 2$. The following theorem is of use when simple graphs with specified values of $\eta$ and $\gamma$ are needed.

*THEOREM 12* [7]: If $G$ is the union of edge-disjoint spanning subgraphs $G_1$ and $G_2$ such that $\eta(G_i) = \gamma(G_i)$ $(i = 1, 2)$, then $\eta(G) = \gamma(G) = \eta(G_1) + \eta(G_2)$.

Thus, given a simple graph $G$ with $\eta(G) = \gamma(G)$, let $H$ be a spanning subgraph of the complement of $G$ such that $\eta(H) = \gamma(H)$ (for example, $H$ can be a spanning tree of $G^C$). Then $\eta(G \cup H) = \gamma(G \cup H) = \eta(G) + \eta(H)$, and $G \cup H$ is a simple graph. This method was used in [8] to prove:

*THEOREM 13* [8]: If $x \geq 1$ is a rational number, then there exists a simple graph $G$ such that $\eta(G) = \gamma(G) = x$.

However, if $x = p/q$ for relatively prime integers $p$ and $q$, the graph $G$ described in the proof of Theorem 13 usually has many more than $p+1$ vertices.

## *THE GENERALIZATION TO MATROIDS*

Certain optimization problems involving edge-toughness and fractional arboricity can be considered best in the context of matroids. For matroids, we shall use the notation of Welsh [33].

Let $M$ be a matroid on $S$ with rank function $\rho$. Define

$$\eta(M) = \min_{T \subset S} \frac{|S-T|}{\rho(S) - \rho(T)}, \tag{3}$$

where the minimum is taken over subsets $T$ of $S$ such that $\rho(T) < \rho(S)$.

Define also

$$\gamma(M) = \max_{\phi \neq T \subseteq S} \frac{|T|}{\rho(T)}. \tag{4}$$

It is easy to check that

$$\eta(M) \le \frac{|S|}{\rho(S)} \le \gamma(M).$$

*THEOREM 14* [6]: If matroid $M$ is the cycle matroid of a graph $G$, then $\gamma(M) = \gamma(G)$ and $\eta(M) = \eta(G)$.

One consequence of this theorem is that questions about the existence of matroids having certain properties of edge-toughness and fractional arboricity can be answered by considering graphs alone. For example:

*THEOREM 11′* : If $x \ge y \ge 1$ are rational numbers, then there is a matroid $M$ with $\gamma(M) = x$ and $\eta(M) = y$.

In matroid theory, there is an easy duality between $\gamma$ and $\eta$. Let $M^*$ stand for the matroid dual of matroid $M$.

*THEOREM 15* [6]: If matroid $M$ is not free, then $\eta(M^*) = \dfrac{\gamma(M)}{\gamma(M)-1}$.

The following theorems generalize Theorems 4 and 5, and for the case $t = 1$, they were obtained by Edmonds [11] (see also [33], Section 8.4). We define a $t$-packing of $M$ to be a family $F$ of bases of $M$ such that each $e \in S$ lies in at most $t$ members of $F$. Let $\eta_t(M)$ be the largest cardinality of a $t$-packing of $M$.

*THEOREM 4′* [10], [6]: For any natural numbers $s$ and $t$, $\eta(M) \geq s/t$ if and only if there is a $t$-packing of $M$ containing $s$ bases.

For a matroid $M$ on a set $S$, we define a $t$-covering of $M$ to be a family $F$ of bases of $M$ such that each $e \in S$ lies in at least $t$ members of $F$. Let $\gamma_t(M)$ be the smallest cardinality of a $t$-covering of $M$.

*THEOREM 5′* [6], [22]: For any natural numbers $s$ and $t$, $\gamma(M) \leq s/t$ if and only if there is a $t$-covering of $M$ containing $s$ bases.

We have the following corollaries to Theorems 4′ and 5′:

*COROLLARY 4′5′A* [6]: Let $M$ be a matroid on $S$ and let $t$ be a natural number. Then $\gamma_t(M) = \lceil t\gamma(M) \rceil$ and $\eta_t(M) = \lfloor t\eta(M) \rfloor$.

*COROLLARY 4′5′B*: Let $M$ be a matroid. Then $\gamma(M) = \min_{1 \leq i \leq \rho(M)} \{\gamma_i(M)/i\}$ and $\eta(M) = \max_{1 \leq i \leq \rho(M)} \{\eta_i(M)/i\}$.

A definition of $\eta$-reduction similar to that for graphs was given for matroids in [6]. An $\eta$-reduced matroid is termed "molecular" by Narayanan and Vartak in [22] and [23] and "irreducible" by Tomizawa in [30]. Using this, we have the following analogue of Theorem 8. As in Theorem 8, parts (d′) and (f′) were shown equivalent in [22].

*THEOREM 8′* [6]: The following are equivalent for a matroid $M$ on a set $S$:

(a′) $|S| = \eta(M)\rho(M)$;

(b′) $|S| = \gamma(M)\rho(M)$;

(c′) $\eta(M) = \gamma(M)$;

(d′) $M$ is $\eta$-reduced;

(e′) there is a function $f : \{1, 2, \ldots, \rho(M)\} \to \mathbf{R}$, where $\mathbf{R}$ is the set of real numbers, such that

(1) $\dfrac{f(r)}{r} \leq \dfrac{f(\rho(M))}{\rho(M)}$ for $1 \leq r \leq \rho(M)$;

(2) $|S| = f(\rho(M))$; and

(3) for any nontrivial submatroid $M'$ of $M$ on subset $S'$ of $S$, $|S'| \leq f(\rho(M'))$;

and

(f′) there is a natural number $t$ and a family $F$ of bases of $M$ such that $F$ is both a $t$-covering and a $t$-packing of $M$.

Let $k$ be a natural number and let $G$ be a graph. Edmonds [12] showed that the edge sets of the subgraphs $H$ of $G$ for which the arboricity of $H$ is at most $k$ are the independent sets of a matroid $M(k)$ on $E(G)$. We have $\eta(H) = \gamma(H) = k$ if and only if $H$ is a union of $k$ edge-disjoint spanning trees. The edge-sets of subgraphs $H$ with $\eta(H) = \gamma(H) = k$ are the bases of $M(k)$. A tempting conjecture is that if $r > 1$ is a rational number, the edge sets of subgraphs $H$ with $\gamma(H) \leq r$ form the independent sets of a matroid. However, Catlin [5] has shown by the following example that this conjecture is false.

Let $G$ be the graph $K_{3,3}$ and let $I = \{A \subseteq E(G) : \gamma(G[A]) \leq 1.49\}$. Let $M$ be $E(G)$ with the subsets of $I$ as independent sets, and suppose $M$ is a matroid. By Theorem 7(a), for any vertex $v$ of $G$, $\gamma(G-v) \geq 6/4 > 1.49$. If $e \in E(G)$, then $\gamma(G-e) \geq 8/5 > 1.49$. By Corollary 6A, if $e_1, e_2 \in E(G)$ and meet at a vertex $v$, then $\gamma(G - \{e_1, e_2\}) = \gamma(G - v) > 1.49$. On the other hand, if $e_1, e_2 \in E(G)$ and are not adjacent, then $\gamma(G - \{e_1, e_2\}) = 1.4 < 1.49$. Thus the bases of $M$ are the edge sets of the subgraphs $G - \{e_1, e_2\}$ for which $e_1$ and $e_2$ do not meet. Hence (by Theorem 1, page 34 of [33]), the bases of the dual $M^*$ of $M$ are the pairs of nonadjacent edges of $G$. Now let $P$ be a path of length three in $G$

with the edges $e_1, e_2, e_3$ being successive along $P$. Then $A = \{e_1, e_3\}$ and $B = \{e_2\}$ are independent in $M^*$ and $|A| = |B| + 1$. By the third axiom of independence, one of $e_1$ or $e_3$ is not adjacent to $e_2$, a contradiction. Thus $M$ is not a matroid.

## *RELATION TO NETWORK DESIGN*

Suppose $w(e)$ is a positive rational number for each edge $e$ of graph $F$ or element $e$ of matroid $F$. Let $d$ be a common denominator for all of the rational numbers $w(e)$, and let $n(e) = dw(e)$ for each edge or element $e$. Define $n(F)$ by replacing each edge or element $e$ of $F$ by $n(e)$ parallel edges or elements. Define $\gamma_w(F)$ and $\eta_w(F)$ by replacing the numerators of (1), (2), (3), and (4) by the sums of the weights of the edges or elements described in those numerators. It is easy to prove:

*THEOREM 16*: If $F$ is a matroid or nontrivial graph, then $\gamma_w(F) = \gamma(n(F))/d$ and $\eta_w(F) = \eta(n(F))/d$.

Since the work described before preserves constant multipliers throughout, the results reported before remain true for graphs or matroids with edges or elements having rational weights. Thus the weighted edges frequently needed by network designers can readily be accomodated by the theory presented so far.

For a graph H, the density $g(H)$ of $H$ is defined in [22] as the ratio $\dfrac{|E(H)|}{|V(H)|-1}$. Clearly, the larger the density of a graph, the more likely it is to be highly connected. But $\gamma(G)$ is the maximum over all nontrivial subgraphs of $G$ of these densities. Thus $\gamma(G)$ is an upper bound on the localized strength of the graph. By part (b) of Theorem 7, this upper bound is the maximum of $\eta(H)$ over the nontrivial subgraphs $H$ of $G$. As observed by Catlin [5], if $\gamma(G)$ is much larger than $\eta(G)$, then there are likely to be many subgraphs $H$ with $\eta(H) > \eta(G)$. But the existence of many proper subgraphs $H$ with high values of $\eta(H)$ will mean that there are comparatively fewer edges of $G$ outside such subgraphs, and so $G$ is weaker outside of those subgraphs. Thus it is preferable to have $\gamma(G) = \eta(G)$

in a network $G$ designed for survivability. In this connection, network designers have long known that maximal planar graphs are superior to other planar graphs for survivability; that $\gamma(G) = \eta(G)$ for such graphs gives further support for this view.

In some cases, a network designer is concerned both with the degradation of communication or transportation that occurs in a long path and with damage to the edges of the network. In this case he would wish to ensure that every edge has a bypass path of length no greater than $k-1$ for some integer $k$; that is, every edge should be in a cycle of length at most $k$. Let $F = \{C_2, C_3, \ldots, C_k\}$. Since $\min_i \eta(C_i) = \frac{k}{k-1}$, by Corollary 10A he must use at least $\frac{k}{k-1}(n-1)$ edges in his network, where $n$ is the number of vertices in the network. This provides a lower bound for the cost of the network if the cost of an edge is fixed.

Another application of edge-toughness and fractional arboricity appears in the design of electrical networks. Kishi and Kajitani [18] defined the unique *principal partition* $G_0, G_1, G_2$ of a graph $G$. They showed that the principal partition can be easily used to determine the minimum number of independent variables, some being currents and some being voltages of edges, needed to do a linear steady-state analysis of an electrical network represented by the graph [32, page 118]. The principal partition itself is readily used to specify the necessary equations for the analysis of the network [26]. These facts are of interest in the present paper because $G_1$ is the maximal subgraph $H$ of $G$ such that $\eta(H) > 2, G_0$ is the maximal subgraph of $G/G_1$ with $\eta(G_0) = 2$, and $G_2$ is $(G/G_1)/G_0$ [6]. The principal partition was refined and extended to matroids by Bruno and Weinberg [3], while a further refinement of the partition was provided by Tomizawa [30]. At about the same time, Narayanan [21] and Narayanan and Vartak [22], [23] gave another view of the same further refinement. Catlin, Hobbs, and Lai [6] describe the connection between edge-toughness, fractional arboricity, and the principal partition in detail, showing that the two functions explicate the refinements of the principal partition.

*ALGORITHMS FOR COMPUTING* $\gamma$ *AND* $\eta$

An algorithm for computing $\eta(G)$ for graph $G$ appears in Cunningham [10]. Catlin, Hobbs, and Lai [6] have shown that Cunningham's algorithm can be modified to compute $\gamma(G)$ as well. However, the algorithm seems to be limited to graphs and graphic matroids, and so an independent algorithm for matroids with a specialization to graphs is presented here.

Given matroids $M_1, \ldots, M_k$, Edmonds [12] provided an algorithm for partitioning the elements of a matroid into disjoint subsets $X_1, \ldots, X_k$ such that $X_i$ is independent in $M_i$ for each $i$. Knuth [19] modified and extended Edmonds' algorithm, and Knuth's version was published by Welsh ([33], pages 365-368). Here we modify Knuth's algorithm to simultaneously compute $\gamma(M)$ and $\eta(M)$ for a matroid $M$.

Let $M$ be a matroid on set $S$, and suppose $|S| = m$. In the following, $k, n_1, \ldots, n_k$, and $v$ are nonnegative integers that are determined by the operation of the algorithm. Here, $k$ is the ordinal label of the presently increasing independent set, $n_1 \ldots, n_k$ are the sizes of the independent sets so far formed, and $v$ is the number of bases among the independent sets found. At Step 4 of the algorithm, $M_t$ is the $t$-parallel extension of $M$ formed by replacing every element of $M$ by $t$ parallel elements as described in [6], and $S_t$ is the set of elements of $M_t$. The steps $(K1)$ through $(K5)$ come from pages 366-367 of [33]. $(K1')$ and $(K5')$ incorporate modifications of the starting and stopping conditions in $(K1)$ and $(K5)$, respectively.

1. Set $t = 1, M_1 = M$, and $S_1 = S$.
2. Set $k = 1$ and $v = 0$.
3. $(K1')$ Start with $X_i = \emptyset$ for all $i$.

   $(K2)$ Construct a digraph $D_t$ on $S_t \cup \{O_1, \ldots, O_k\}$ as follows: If $x \in S_t$ is not in $X_j$ and if $X_j \cup \{x\}$ is independent in $M_t$, $1 \leq j \leq k$, put arc $(x, O_j)$ in $D_t$. If $x \in S_t - X_j$ and if $X_j \cup \{x\}$ is dependent, let $C$ be the unique circuit in $X_j \cup \{x\}$. Put $(x, y)$ in $D_t$ for each $y \neq x$ in $C$.

($K3$) Attempt to find a $y \in S_t - \bigcup X_i$ such that there is a dipath of minimal length from $y$ to $O_k$ in $D_t$.

($K4$) If such a path $y = x_0, x_1, x_2, \ldots, x_p, O_k$ exists, for each $j = 1, 2, \ldots, p$, successively let $i_j$ be the subscript such that $x_j \in X_{i_j}$ and reform $X_{i_j}$ as $(X_{i_j} \cup \{x_{j-1}\}) \backslash x_j$. Last, reform $X_k$ as $X_k \cup \{x_p\}$. Go to ($K2$).

($K5'$) If no such $y$ exists, $X_k$ is as large as possible. Set $n_k = |X_k|$. If $|X_k| = \rho S$, increase $v$ by 1. If $S_t - \bigcup X_i \neq \emptyset$, increase $k$ by 1 and go to ($K2$).

4. Now $S_t - \bigcup X_i = \emptyset$; set $r_t = v$ and $s_t = k$. If $r_t = s_t$, then $\eta(M) = \gamma(M) = r_t/t$; stop. Otherwise, if $t < \rho S$, increase $t$ by one, form $M_t$ and $S_t$, and go to Step 2.

5. Then $\eta(M) = \max\{r_i/i\}$ and $\gamma(M) = \min\{s_i/i\}$

*PROOF* : Note first that the modifications (from ($K1$)-($K5$) to ($K1'$), ($K2$)-($K4$), ($K5'$)) to Knuth's algorithm, hereafter denoted ($K1$)-($K5$), change only the initialization criterion (to make the number of sets involved indefinite) and the stopping criterion. Thus if a fixed number of sets $X_i$ and fixed sizes $n_i$ for them were given to the above algorithm, it would work exactly when ($K1$)-($K5$) works. Also, ($K1$)-($K5$) first fills $X_2$ to $n_2$, then ..., then fills $X_{k-1}$ to $n_{k-1}$. That is, it does not start putting elements in $X_j$ until $X_1$, ..., $X_{j-1}$ are at sizes $n_1, \ldots, n_{j-1}$, respectively.

Fix $t$. Suppose in $S_t$ there is a maximum packing with $\alpha$ distinct bases of $M_t$. Let $n_1 = n_2 = \ldots = n_\alpha = \rho S_t = \rho S$. Applying ($K1$)-($K5$) to $S_t$, by Theorem 1, page 367 of [33], a packing of $\alpha$ disjoint bases is found. But this same algorithm is used above, so $r_t \geq \alpha$. On the other hand, since $\alpha$ is the largest number of disjoint bases of $M_t$ in $S_t$, then $r_t \leq \alpha$. Thus $r_t = \alpha$.

With the same $t$, next suppose there is in $S_t$ a minimum covering with $\beta$ disjoint independent sets. Letting $n_1, \ldots, n_\beta$ be the sizes of these independent sets, choose the covering $(Y_1, \ldots, Y_\beta)$ among all minimum coverings so that $n_1$ is as large as possible, and then $n_2$ is as large as possible, and then ..., and then $n_{\beta-1}$ is as large as possible; thus

$n_1 \geq n_2 \geq \ldots \geq n_\beta$. Suppose for some $i < j$ that $|Y_i| < \rho S$ and there is a $y \in Y_j$ such that $Y_i \cup \{y\}$ is independent. But then forming $(Y_1, \ldots, Y_{i-1}, Y_i \cup \{y\}, Y_{i+1}, \ldots, Y_{j-1}, Y_j - \{y\}, Y_{j+1}, \ldots, Y_\beta)$ produces a covering which is contrary to our choice of $(Y_1, \ldots, Y_\beta)$. Hence $(K1)$-$(K5)$ finds a disjoint cover $(X_1, \ldots, X_\beta)$ of $S_t$ with $|X_i| = n_i$. Thus $s_t \leq \beta$. But $s_t \geq \beta$ since $(X_1, \ldots, X_{s_t})$ is a cover and $(Y_1, \ldots, Y_\beta)$ is a minimal cover. Thus $s_t = \beta$.

The stopping criterion in Step 4 is just Part (f′) of Theorem 8′.

For Step 5, we notice that a 1-packing or 1-covering of $M_t$ is a $t$-packing or a $t$-covering of $M$. By definition of $\gamma_t$ and $\eta_t$ and of $r_t$ and $s_t$, we have $r_t = \gamma_t$ and $s_t = \gamma_t$ for each value of $t$. Thus the conclusion of Step 5 follows from Corollary $4'5'B$. ∎

The Knuth algorithm requires at most $m^3 + m^2k$ tests of independence in the matroids $M_i$ ([33], page 368). Although $k$ in our algorithm is not a constant, we do have $k \leq mt$. Thus in $M_t$ we use no more than $cm^3t^3$ tests of independence. Since the denominator of each of $\gamma(M)$ and $\eta(M)$ is at most $\rho S$, our algorithm requires at most $\sum_{t=1}^{\rho S} cm^3t^3$ tests of independence in the matroid $M$. Thus the above algorithm is $O(m^3(\rho S)^4)$ in its number of tests of independence.

Next is the graph algorithm for computing $\eta$ and $\gamma$. Let $G$ be a graph, and suppose $|E(G)| = m$ and $|V(G)| = n$. In the following, $k$, $n_1, \ldots, n_k$, and $v$ are positive integers that are determined by the operation of the algorithm. Here, $k$ is the ordinal label of the presently increasing forest, $n_1, \ldots, n_k$ are the sizes of the forests so far formed, and $v$ is the number of spanning trees among the forests found. At Step 4 of the algorithm, $G_t$ is formed from $G$ by replacing every edge of $G$ by $t$ parallel edges. The steps $(K1)$ through $(K5)$ and $(K1')$ and $(K5')$ are as before.

1. Set $t = 1$. Let $G_1 = G$.

2. Set $k = 1$ and $v = 0$.

3. $(K1')$ Start with $X_i = \emptyset$ for all $i$.

   $(K2)$ Construct a digraph $D_t$ on $E(G_t) \cup \{O_1, \ldots, O_k\}$ as follows: If $x \in E(G_t)$

is not in $X_j$ and if $X_j \cup \{x\}$ is a forest in $G_t$, for $1 \leq j \leq k$, put arc $(x, O_j)$ in $D_t$. If $x \in E(G_t) - X_j$ and if $X_j \cup \{x\}$ is not a forest, let $C$ be the unique cycle in $X_j \cup \{x\}$. Put $(x, y)$ in $D_t$ for each $y \neq x$ in $C$.

($K3$) Attempt to find a $y \in E(G_t) - \bigcup X_i$ such that there is a dipath of minimal length from $y$ to $O_k$ in $D_t$.

($K4$) If such a path $y = x_0, x_1, x_2, \ldots, x_p, O_k$ exists, for each $j = 1, 2, \ldots, p$, successively let $i_j$ be the subscript such that $x_j \in X_{i_j}$ and reform $X_{i_j}$ as $(X_{i_j} \cup \{x_{j-1}\}) \backslash x_j$. Last, reform $X_k$ as $X_k \cup \{x_p\}$. Go to ($K2$).

($K5'$) If no such $y$ exists, $X_k$ is as large as possible. Set $n_k = |X_k|$. If $|X_k| = n - 1$, increase $v$ by 1. If $E(G_t) - \bigcup X_i \neq \emptyset$; increase $k$ by 1 and go to ($K2$).

4. Now $E(G_t) - \bigcup X_i = \emptyset$; set $r_t = v$ and $s_t = k$. If $r_t = s_t$, then $\eta(G) = \gamma(G) = r_t/t$; stop. Otherwise, if $t < n - 1$, increase $t$ by one, form $G_t$ by replacing each edge of $G$ by $t$ parallel edges, and go to Step 2.

5. Then $\eta(G) = \max\{r_i/i\}$ and $\gamma(G) = \min\{s_i/i\}$.

*PROOF:* This is just the algorithm for the cycle matroid of a graph translated into the terminology of graph theory. ∎

As with the matroid algorithm, this algorithm is $O(m^3n^4)$ in tests for cycles.

*ACKNOWLEDGEMENT*

My thanks are due to P. A. Catlin and H.-J. Lai of Wayne State University, Detroit, Michigan, and to J. W. Grossman of Oakland University, Rochester, Michigan, for many helpful discussions during the preparation of this paper.

## REFERENCES

1. B. Bollobás, Cycles and semi-topological configurations, *Theory and Applications of Graphs* (Kalamazoo, Mich., 1976), 66-74, *Lecture Notes in Math.* 642, Springer-Verlag, Berlin, 1978.

2. J. A. Bondy and U. S. R. Murty, *Graph Theory with Applications*, American Elsevier Publ. Co., Inc., 1976.

3. J. Bruno and L. Weinberg, The principal minors of a matroid, *Linear Algebra and its Applications* 4 (1971), 17-54.

4. P. A. Catlin, The reduction of graph families closed under contraction, preprint.

5. P. A. Catlin, Private communication.

6. P. A. Catlin, A. M. Hobbs, and H. Lai, Matroid Unions, edge-toughness, and fractional arboricity, preprint.

7. P. A. Catlin, K. C. Foster, J. W. Grossman, and A. M. Hobbs, Graphs with specified edge-toughness and fractional arboricity, preprint.

8. P.A. Catlin, J. W. Grossman, and A.M. Hobbs, Graphs with uniform density, preprint.

9. V. Chvátal, Tough graphs and Hamiltonian circuits, *Discrete Math.* 5 (1973), 215-228.

10. W. H. Cunningham, Optimal attack and reinforcement of a network, *J. Assoc. Comp. Mach.* 32 (1985), 549-561.

11. J. Edmonds, Lehman's switching game and a theorem of Tutte and Nash-Williams, *J. Res. Nat. Bur. Stand.* 69B (1965), 73-77.

12. J. Edmonds, Matroid partition, *Lectures in Appl. Math.* 11 *(Mathematics of the Decision Sciences)* (1967), 335-346.

13. P. Erdös, Elementary Problem E3255, *American Math. Monthly* 95(1988), 259.

14. A. Frank, Covering branchings, *Acta Scientiarum Math.* 41 (1979), 77-81.

15. A. Frank, On the orientation of graphs, *J. Combin. Theory (B)* 28 (1980), 251-261.

16. C. Greene and T. L. Magnanti, Some abstract pivot algorithms, *SIAM J. Appl. Math.* 29 (1975), 530-539.

17. D. Gusfield, Connectivity and edge-disjoint spanning trees, *Information Processing Letters* 16 (1983), 87-89.

18. G. Kishi and Y. Kajitani, Maximally distinct trees in a linear graph, *Electronics and Communications in Japan* 51-A, No. 5 (1968), 35-42, translated from *Trans. Inst. Elect. Commun. Engin. Japan* J51-A (1968), 196-204.

19. D. E. Knuth, Matroid partitioning (preprint), Stanford University STAN-CS-73-342 (1973), 1-12.

20. S. Kundu, Bounds on the number of edge-disjoint spanning trees, *J. Combin. Theory (B)* 17 (1974), 199-203.

21. H. Narayanan, Theory of matroids and network analysis, Ph.D. Thesis submitted to the Department of Electrical Engineering, Indian Institute of Technology, Bombay, Febraury, 1974.

22. H. Narayanan and M. N. Vartak, On molecular and atomic matroids, *Combinatorics and Graph Theory* (Calcutta, 1980), 358-364, Lecture Notes in Mathematics 885, Springer-Verlag, Berlin, 1981.

23. H. Narayanan and N. Vartak, An elementary approach to the principal partition of a matroid, *Trans. IECE Japan* E-64 (1981), 227-234.

24. C. St. J. A. Nash-Williams, Edge-disjoint spanning trees of finite graphs, *J. London Math. Soc.* 36 (1961), 445-450.

25. C. St. J. A. Nash-Williams, Decomposition of finite graphs into forests, *J. London Math. Soc.* 39 (1964), 12.

26. T. Ohtsuki, Y. Ishizaki, and H. Watanabe, Network Analysis and topological degrees of freedom, *Electronics and Communications in Japan* 51-A, No. 6 (1968), 33-40, translated from *Trans. Inst. Elect. Commun. Engin. Japan* J51-A (1968), 238-246.

27. J. G. Oxley, On a packing problem for infinite graphs and independence spaces. *J. Combin. Theory. (B)* 26 (1979), 123-180.

28. C. Palm and J. M. S. Simoes-Pereira, Edge sets of hypergraphs with a $B_q$- like property and partition numbers of graphs, *Colloquia Math. Soc. J. Bolyai* 18, *Combinatorics*, Vol. 2, (Keszthely, Hungary, 1976), 793-804, North-Holland Publ. Co., Amsterdam, 1978.

29. R. E. Tarjan, Edge-disjoint spanning trees and depth-first search, *Acta Informatica* 6 (1976), 171-185.

30. N. Tomizawa, Strongly irreducible matroids and principal partition of a matroid into strongly irreducible minors, *Electronics and Communications in Japan* J59-A, No. 2 (1976), 1-10, translated from *Trans. Elect. Comm. Engin. Japan* J59-A, No. 2 (1976), 83-92.

31. W. T. Tutte, On the problem of decomposing a graph into $n$ connected factors, *J.*

*London Math. Soc.* 36 (1961), 221-230.

32. H. Watanabe, Activities on circuit theory in Japan, *IEEE Trans. Circuits and Systems* CAS-31 (1984), 112-123.

33. D. J. A. Welsh, *Matroid Theory*, Academic Press, London, 1976.

Contemporary Mathematics
Volume 89, 1989

# Directed Graphs and the Compaction of IC Designs

by Bradley W. Jackson

San Jose State University

San Jose, CA. 95192

The layout of an integrated circuit consists of the placement of the cells composing the design and the routing of the necessary interconnections between these cells. The goal of this process is to complete the placement and routing using the minimum possible area so that a collection of design rules depending on the technology are satisfied. With the increasing complexity of VLSI computer aided layout design has become necessary to cut the amount of time needed to complete a design and to eliminate human errors. However the present computer aided design tools for placement and routing tend to use more area than hand layout. Thus it has become desirable to include a final step in the layout design process, layout spacing and compaction. The placement and routing steps give an initial layout, and the aim of layout compaction is to obtain a final layout using minimum area without changing the topology of the initial layout (see [6]).

To reduce the complexity of the design process we use a symbolic layout which represents the wires and cells of an integrated circuit by lines and rectangles in a coordinate plane. There exist many techniques for analyzing the simplest designs which contain only vertical and horizontal lines, and design rules that can be expressed using constraints on the distances between pairs of parallel lines. The aim of this paper is to discuss techniques for analyzing more complicated designs that have 45 degree lines and contain more complex constraints involving symmetry. The research described in this paper is partially supported by IBM and Intel as part of an applied mathematics and computer science project at San Jose State University.

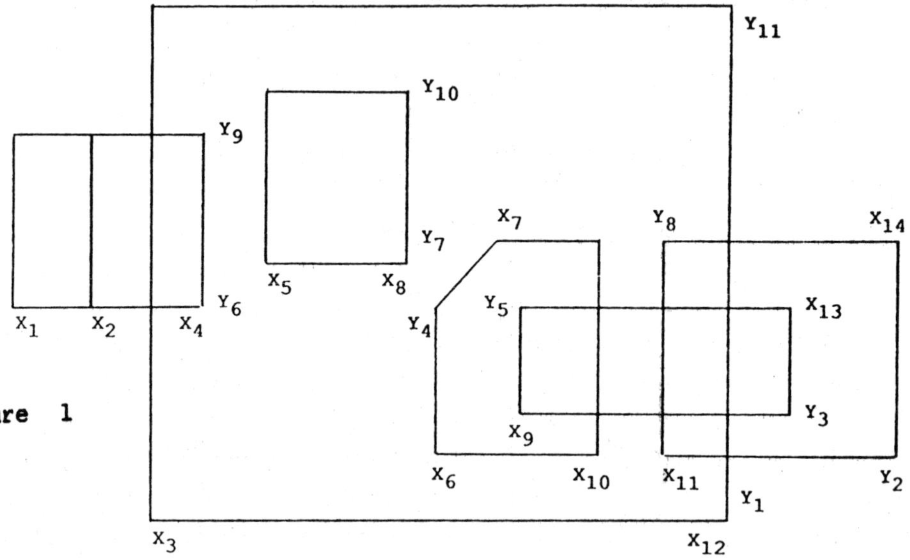


**Figure 1**

To describe the positions of the lines in an integrated circuit layout we assume that it is drawn in a coordinate plane as in figure 1. A feasible solution of the circuit design problem will be a set of nonnegative integer values for the endpoints of these line segments which satisfy the constraints given by the design rules. An optimal solution will be a feasible solution for which the overall area is minimized. The position of any vertical line in the design can be described by a single x-coordinate and likewise the position of any horizontal line can be described by a single y-coordinate. For a design that is only subject to basic separation constraints the x-values and the y-values are independent so we can compact the area by first minimizing the x-values and then the y-values.

Consider the problem of compacting an integrated circuit design in the x-direction subject to basic separation constraints on the distances between pairs of vertical lines. This can be described as an integer programming problem. The variables will be the x-coordinates of the vertical lines in the layout of the

IC design, which we denote by $x_1, x_2, \ldots, x_n$ and one other variable $x$. The variable $x$ will be equal to $\max\{x_1, x_2, \ldots, x_n\}$ and we assume that each variable is a nonegative integer so the constraints $0 \leq x_i \leq x$ are added. In addition, three kinds of separation constraints can occur. We have minimum distance constraints of the form $x_i - x_j \geq c_{ij} > 0$. These are often used to enforce minimum separation requirements needed by the chip fabrication process. There are also maximum distance constraints of the form $x_i - x_j \leq c_{ij}$. These may occur in a design when two cells are required to be close enough to avoid wire delay problems. Finally we have equality constraints of the form $x_i - x_j = c_{ij}$. These constraints can be used to ensure the precise alignment of two cells in the design. Our goal is to find a set of nonnegative integer values which satisfy the constraints and minimize the value of the objective function $x$. In general, integer programming is NP-complete so it is unlikely that any efficient algorithm which handles all such problems will ever be found. Thus for any subclass of integer programming problems one should hope to find special techniques for solving them.

One basic algorithm finds a solution efficiently by modeling this problem using a directed graph. First we express all of the constraints using greater than or equal to inequalities. Less than or equal to constraints $x_i - x_j \leq k$ are transformed into greater than or equal to constraints with a negative constraint value $x_j - x_i \geq -k$ while equality constraints can be represented by two inequalities. We consider the graph with one vertex for each variable $x_i$ and one directed edge from $x_j$ to $x_i$ with weight $k$ for each constraint of the form $x_i - x_j \geq k$. Figure 2 gives an example of a set of constraints and the corresponding constraint graph.

**Figure 2**

$x_2 - x_1 \geq 2$

$x_3 - x_2 = 10$

$x_3 - x_1 \geq 13$

$x_4 - x_3 \geq 1$

$x_4 - x_2 \leq 12$

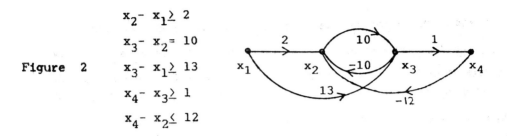

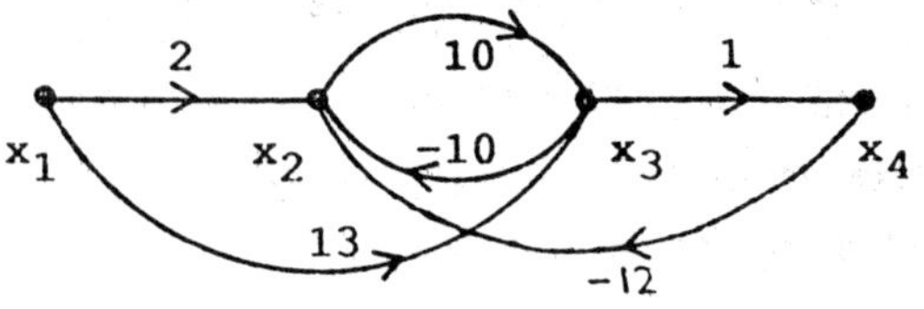


Other relationships between the x-coordinates can be determined by looking at the geometry of the constraint graph. We say that the weight of a directed path or a directed cycle is equal to the sum of the weights of its edges. If there is a directed path P from vertex $x_j$ to vertex $x_i$, then one can easily prove by induction that $x_i - x_j \geq$ weight of P. Similarly if there is a directed cycle C in the constraint graph containing a vertex $x_i$ we see that $x_i - x_i \geq$ weight of C. This last statement shows that a set of constraints will not have a feasible solution if the constraint graph has a positive weight cycle. In fact, if the constraint graph has no positive weight cycles, then there is a convenient description of an optimal solution of this one-dimensional compaction. We obtain an optimal solution by setting the value of $x_i$ equal to the maximum weight of a directed path in the constraint graph that ends at $x_i$. Equivalently we can define an initial vertex 0 whose value is defined to be 0 and set $x_i$ equal to the maximum weight of a path starting at 0 and ending at $x_i$ as in [2,4]. The maximum weight path solution for the example in figure 2 is thus, $x_1 = 0$, $x_2 = 3$, $x_3 = 13$, $x_4 = 14$.

The algorithm we use to obtain the maximum path solution is an adaptation of Ford's algorithm for determining minimum weight paths in a directed graph. A solution is obtained by repeatedly processing the entire list of constraints. If no changes are made in the ith pass through the constraints, then a solution has been obtained. If changes are still made in the nth pass then a positive weight cycle can be found by backtracking. In either case the algorithm can be terminated after at most n passes through the constraint list, so the worst case efficiency of this algorithm is O(nm) where n is the number of vertices in the constraint graph and m is the number of directed edges.

For a typical design problem the efficiency of this algorithm can be increased by breaking a graph into its strongly connected components as explained in [5] for algorithms that find minimum weight paths in a directed graph. A strongly connected component in a constraint graph consists of a maximal subset of vertices in which every pair of vertices is joined by a directed path and the directed edges that connect them. By performing a depth first search on the constraint graph we can determine its SCC's and arrange them in order so that each SCC has no incoming edges when the preceding SCC's are removed (see [3]). We start with an SCC that has no incoming edges and repeatedly process the constraints inside this SCC until the values for these vertices have been established (or a positive weight cycle has been discovered). The edges leaving this SCC are processed one time and then the vertices of this component are removed from the constraint graph. The algorithm continues processing one component at a time until all of the SCC's have been processed in this manner. The worst case efficiency of this algorithm is $O(m + \sum n_i m_i)$ where the number of vertices in the ith SCC is $n_i$ and the number of edges in the ith SCC is $m_i$. In most design problems that arise there will be few maximum distance constraints and the SCC's are likely to be small so the average complexity for this version of the algorithm is probably close to O(m).

## The Basic Algorithm

The problem is to compute the minimum values for $x_1, x_2,$

... $,x_n$ subject to constraints of the form $x_i - x_j \geq c_{ij}$, $x_i - x_j = c_{ij}$, or $x_i - x_j \leq c_{ij}$. The solution to the problem requires the following three steps.

1) Initialization

Read the number of vertices. Read the constraints and change them all to the form $x_i - x_j \geq c_{ij}$. Store the constraints incident to each vertex in a list.

For each vertex $x_i$, set its value $V(x_i)$ equal to 0.

2. Find the strongly connected components of the constraint graph using a depth first search.

3. We start with the inequalities incident to the vertices of a strongly connected "source" component, that is one with no incoming edges. We process the inequality constraints inside a strongly connected component C repeatedly as described above. To process the inequality $x_i - x_j \geq c_{ij}$ we set $V(x_i) = \max\{V(x_i), V(x_j)+c_{ij}\}$. The values of the vertices in C have been determined when no changes have been made after an entire pass through the constraints. If the component C has p vertices, then a positive weight cycle has been detected when changes are still made on the pth pass through the constraints, and in this case there is no solution. Continue by processing the constraints which start in C and end in some other component once. Since the vertices in C have now received their minimum values we remove these vertices and proceed to the next strongly connected source component until all vertices $x_i$ have received their minimum values. This is the final solution.

We now illustrate the algorithm on a small test problem defined by the constraint graph in figure 3.

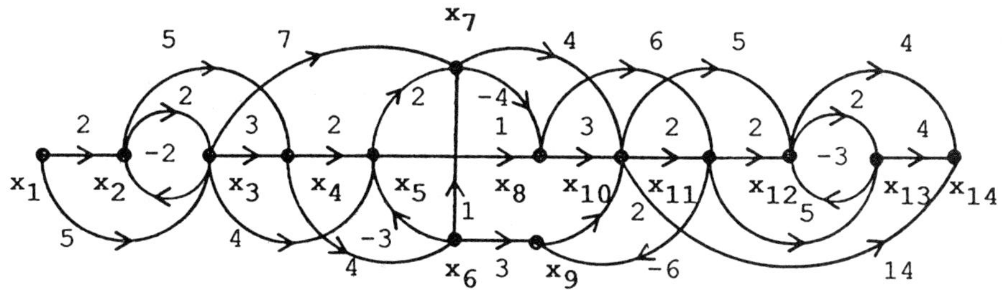


**Figure 3**

1. Initialization

There are 14 variables and 30 constraints. The constraints given in "greater than or equal to" form are listed for each vertex in figure 4.

$x_1$; $x_2 - x_1 \geq 2$, $x_3 - x_1 \geq 5$

$x_2$; $x_3 - x_2 \geq 2$, $x_4 - x_2 \geq 5$

$x_3$; $x_2 - x_3 \geq -2$, $x_4 - x_3 \geq 3$, $x_5 - x_3 \geq 4$, $x_7 - x_3 \geq 7$

$x_4$; $x_5 - x_4 \geq 2$, $x_6 - x_4 \geq 4$

$x_5$; $x_7 - x_5 \geq 2$, $x_8 - x_5 \geq 1$

$x_6$; $x_5 - x_6 \geq -3$, $x_7 - x_6 \geq 1$, $x_9 - x_6 \geq 3$

$x_7$; $x_8 - x_7 \geq -4$, $x_{10} - x_7 \geq 4$

$x_8$; $x_{10} - x_8 \geq 3$, $x_{11} - x_8 \geq 6$

$x_9$; $x_{10} - x_9 \geq 2$

$x_{10}$; $x_{11} - x_{10} \geq 2$, $x_{12} - x_{10} \geq 5$, $x_{14} - x_{10} \geq 14$

$x_{11}$; $x_9 - x_{11} \geq -6$, $x_{12} - x_{11} \geq 2$, $x_{13} - x_{11} \geq 5$

$x_{12}$; $x_{13} - x_{12} \geq 2$, $x_{14} - x_{12} \geq 4$

$x_{13}$; $x_{12} - x_{13} \geq -3$, $x_{14} - x_{13} \geq 4$

$x_{14}$; empty

**Figure 4**

**Constraint Lists**

2. Depth First Search

The strongly connected components determined by the depth first search are listed below.

$C_1=\{x_1\}$, $C_2=\{x_2,x_3\}$, $C_3=\{x_4\}$, $C_4=\{x_5\}$, $C_5=\{x_6\}$, $C_6=\{x_7\}$, $C_7=\{x_8\}$, $C_8=\{x_9,x_{10},x_{11}\}$, $C_9=\{x_{12},x_{13}\}$, $C_{10}=\{x_{14}\}$

3. Processing Constraints

Each value is initially zero and the final values of the variables assigned during the processing of the inequality constraints are listed below.

$x_1 = 0$, $x_2 = 3$, $x_3 = 5$, $x_4 = 8$, $x_5 = 10$, $x_6 = 12$, $x_7 = 13$, $x_8 = 11$, $x_9 = 15$, $x_{10} = 17$, $x_{11} = 19$, $x_{12} = 22$, $x_{13} = 24$, $x_{14} = 31$.

**Symmetry Constraints**

To produce electrically balanced circuits engineers would like to use design rules which specify symmetry in certain parts of the integrated circuit design. We consider two types of symmetry. The first example is where pairs of lines are equally spaced about an axis with coordinate $x_i$. If lines $x_h$ and $x_j$, $x_g$ and $x_k$, etc. are equally spaced about $x_i$ as in figure 5 then we introduce the constraints $x_j - x_i = x_i - x_h$, $x_k - x_i = x_i - x_g$, ... .

**Figure 5**

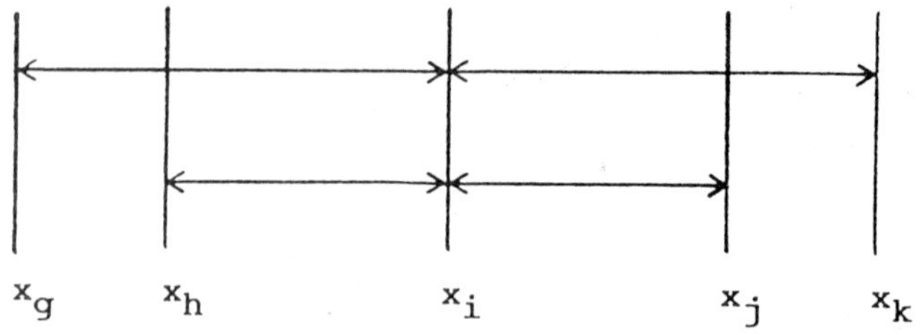

These are called 3-variable symmetry constraints. In order to continue using the directed graph model we replace each 3-variable symmetry constraint by two 2-variable constraints which have a constraint value that is a variable rather than a constant. The constraint $x_j - x_i = x_i - x_h$ is replaced by two constraints $x_j - x_i = a$ and $x_i - x_h = a$, the constraint $x_k - x_i = x_i - x_g$ is replaced by two constraints $x_k - x_i = b$ and $x_i - x_g = b$, and so on. Thus we now have variable weight edges in the constraint graph. Figure 6 shows how these constraints would appear in the constraint graph.

**Figure 6**

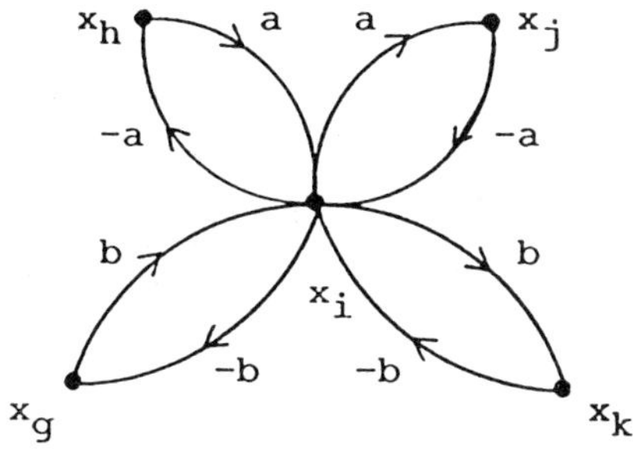


We might also like to have the lines in two (or more) different parts of the design equally spaced about separate axes $x_i$ and $x_p$ as in figure 7.

**Figure 7**

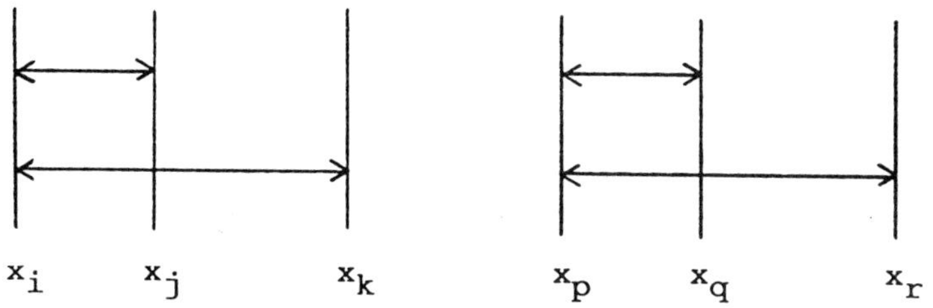


To describe this we introduce 4-variable symmetry constraints of the form $x_j - x_i = x_q - x_p$, $x_k - x_i = x_r - x_p$, etc. In a similar manner each 4-variable symmetry constraint is replaced by two 2-variable constraints with variable constraint values. Therefore the constraint $x_j - x_i = x_q - x_p$ is replaced by $x_j - x_i = c$ and $x_q - x_p = c$, and the constraint $x_k - x_i = x_r - x_p$ is replaced by $x_k - x_i =$

d and $x_r - x_p = d$. Figure 8 illustrates how these variable weight edges would appear in the constraint graph.

Figure 8

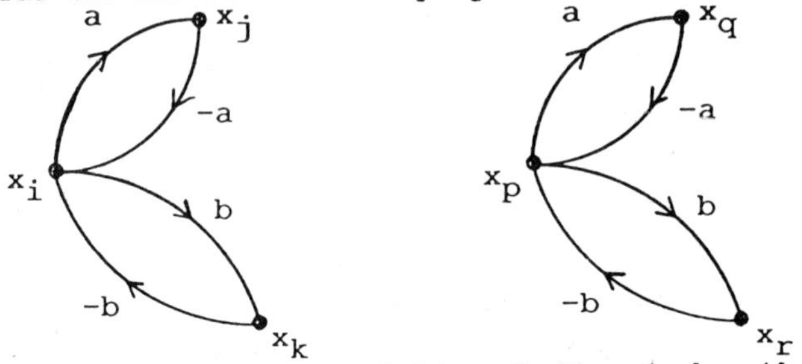


In the constraint graph the weights of directed paths and directed cycles are now represented by linear functions of the variable edge weights rather than simple numerical quantities. To solve problems involving symmetry constraints our one-dimensional compaction can now be performed in two steps each of which can be thought of as an integer programming problem.

1) Analyze the cycles of the constraint graph and select numerical values for the variable edge weights so that no positive weight cycles are introduced.

2) Solve the resulting constraints using the basic algorithm.

Since the second problem is familiar we concentrate on the first problem which we refer to as the auxiliary problem. The auxiliary problem can be thought of as an integer programming problem involving the variables edge weights a,b,c, ... . Let $c_1(a,b,c, \ldots)$, $c_2(a,b,c, \ldots)$, $c_3(a,b,c, \ldots)$, ..., be the weights of the directed cycles in the constraint graph, and let $p_1(a,b,c, \ldots)$, $p_2(a,b,c, \ldots)$, ..., be the weights of the directed paths. The directed cycles must not have positive weight so the cycles in the constraint graph can be analyzed to find constraints on these variable edge weights. Any solution to the original problem will give us integer values for a,b,c, ... which satisfy $c_1(a,b,c, \ldots) \leq 0$, $c_2(a,b,c, \ldots) \leq 0$, $c_3(a,b,c,$

... ) $\leq$ 0, ... . Again the overall maximum path length is to be minimized, so we introduce a new variable P satisfying the constraints $P \geq p_i(a,b,c, \ldots)$. Thus we would like to find integer values of a,b,c, ... that satisfy all of the constraints while minimizing the objective function P. Since the constraints on a,b,c, ... can be arbitrary linear functions in general, it is known that even finding a feasible solution to these problems is NP-complete. Thus to solve these problems efficiently we should expect to restrict our attention to a subclass of the problems.

Each of the two kinds of symmetry constraints that we have mentioned above introduces a set of variable weight edges into the constraint graph which are incident to one or more vertices called axes. It is again appropriate to consider the strongly connected components of the constraint graph. We will look at problems where no more than one axis of symmetry is contained in any strongly connected component. An axis may be incident to arbitrarily many variable weight edges, but any cycle passing through the axis will contain at most two of these variable weight edges. Since a cycle is always contained entirely in one component, any cycle will contain at most two directed edges of variable weight so the corresponding constraints on variable edge weights will contain no more than two variables. In addition, the coefficients of the variables in these constraints are either -1, 0, or 1. Thus we can have any of the following three types of constraints on variable edge weights.

$$i_1 \leq a \leq i_2,\ i_3 \leq b \leq i_4,\ \ldots$$

$$j_1 \leq b-a \leq j_2,\ j_3 \leq c-b \leq j_4,\ \ldots$$

$$k_1 \leq a+b \leq k_2,\ k_3 \leq b+c \leq k_4,\ \ldots$$

If only the first two types of constraints are present, then our basic algorithm can be used to solve the auxiliary problem. This will be the case in any strongly connected component that contains a single axis of 4-variable symmetry. If the third type of constraint is present, then a different technique is needed. This will be the case in any strongly connected component that contains a single axis of 3-variable symmetry. It turns out that these integer programming problems can also be solved efficiently using directed graph techniques. In this case we adapt a method of R. Shostak and B. Aspvall, Y. Shiloach [1,7] for solving linear programming problems with 2-variable constraints.

The second generation algorithm is able to handle some additional constraints in an IC design. For problems involving symmetry it is probably a reasonable restriction to assume that there is at most one axis of symmetry in any strongly connected component. If the constraint graph contains at most one axis of symmetry in any strongly connected component, this algorithm can always find a feasible solution. The solution obtained will be in some sense locally optimal on strongly connected components, but we cannot guarantee that the final solution is globally optimal. However we do know that a feasible solution will be found if one exists. The worst case complexity of the basic algorithm is $O(nm)$ for a constraint graph with n vertices and m edges but the complexity of the second generation algorithm is increased to $O(n^2m)$.

The second generation algorithm must first find the cycles in the constraint graph that contain variable weight edges and then analyze the corresponding constraints to obtain feasible values for the variable edge weights. In a strongly connected component containing a single axis of symmetry a cycle will consist of two (or fewer) variable weight edges followed by a directed path of numerical weight edges. It may be helpful to think of any axis of symmetry in a strongly connected component being labeled by 0. In a similar way the terminal vertex of any edge of variable weight $\pm a$ which starts at 0 can be labeled by its variable edge weight $\pm a$. Any path of weight $w_{ab}$ joining vertex a to vertex b will mean that $b - a \geq w_{ab}$. Thus to obtain the constraints on the variable edge weights a,b,c, ... , it will suffice to find the maximum weight numerical path between any two vertices that are the endpoints of a variable weight edge. Since the weight of any cycle must be nonpositive this information is used to find the constraints on the variable edge weights a,b,c, ... . The difficulty of finding a set of feasible values for the edge weights a,b,c, ... , depends on the kind of symmetry that is required. We have the following description of the second generation algorithm.

**The Second Generation Algorithm**

1. Read in the basic constraints. Process defined values by adding equalities between pairs of defined vertices. Read in the symmetry constraints. Process symmetry constraints by adding equalities with variable constraint values, and by adding the corresponding variable weight edges to the constraint graph.

2. Perform a depth first search to find the strongly connected components when both numerical and variable weight edges are included in the constraint graph. Check that each strongly connected component contains at most one axis.

3. a) In components containing only basic constraints, the constraints will be processed repeatedly as in our basic algorithm. If the ith component has $n_i$ vertices and $m_i$ edges, then in this step the number of operations required to find the values of the vertices is $O(n_i m_i)$.

b) For a component that contains a single axis of symmetry we first find the constraints on the variable edge weights. To do this we use an adaptation of Floyd's algorithm which finds the maximum weight path between every pair of vertices in a directed graph. A maximum path of weight $w_{ab}$ joining vertex a to vertex b gives us the constraint $b - a \geq w_{ab}$. the number of operations required for this computation is $O(n_i^3)$. To solve the auxiliary problem for this component we store the constraints on variable weight edges in a separate variable constraint graph.

i) To process a component containing an axis of 4-variable symmetry we use a constraint graph with vertices 0,a,b, ... . As before each constraint is represented by a single directed edge. (Note that two or more components contain variable weight edges a,b, ... so we must solve these auxiliary problems simultaneously to ensure that the strongest constraints on the variables are used.) The minimum values of the variables can be found using a slight modification of our basic algorithm on this new constraint graph to compute the maximum weight path from the axis 0 to every other vertex. There is no feasible solution if and only if the constraint graph has a positive weight cycle. In this case the number of operations required is $O(n_i m_i)$.

ii) To process a component containing an axis of 3-variable symmetry we use a constraint graph with vertices ...,-b,-a,0,a,b, ... . Each constraint is now represented by a pair of directed edges. For example, a constraint $a - b \geq k_1$ (equivalently $-b - (-a) \geq k_1$) is represented by a directed edge of weight $k_1$ from b to a and another directed edge of weight $k_1$ from -a to -b. (The strongest constraints on the variables are again obtained by looking at a pair of maximum path lengths in the constraint graph.) In this case the auxiliary problem is complicated by the addition of a new type of 2-variable constraint, so we adapt the general method described by R. Shostak for solving linear programming problems with arbitrary 2-variable constraints. For each variable a we first find the maximum path length from vertex -a to vertex a (and from a to -a). If the weight of a maximum path from -a to a is $w_a$ add edges of weight $[w_a/2]$ from -a to 0 and from 0 to a. According to Shostak [7], this addition ensures that the problem has a feasible solution if and only if the constraint graph has no positive weight cycle. In this case however, the minimum values cannot always be determined simultaneously. We proceed by finding the maximum path length from vertex 0 to every other vertex. Fix the value of some undefined variable. Add the corresponding equality to the constraint graph and continue by finding the maximum path length from 0 to every other vertex. Repeat these steps until each variable has been assigned a value. Since the values of the variables are computed successivley in this case the worst case complexity is increased to $O(n_i^2 m_i)$. In practice the successive calculations can be started from the previously computed values and can be computed after just a few iterations of the basic algorithm. Thus the average number of operations needed to deal with an axis of 3-variable symmetry will not be much greater than the number needed to deal with a 4-variable axis of symmetry.

4. Replace any variable weight edges in the original constraint graph by numerical weight edges. Each variable receives the numerical value just computed in step 4. Compute the minimum values in the resulting constraint graph using the basic algorithm to obtain the final solution.

To illustrate the second generation algorithm consider the original list of constraints in figure 4 with the following 4-variable symmetry constraints added. Suppose that $x_1$ and $x_{11}$ are axes of symmetry with $x_2 - x_1 = x_{12} - x_{11}$, $x_3 - x_1 = x_{13} - x_{11}$, and $x_4 - x_1 = x_{14} - x_{11}$. In the first step we add the following equalities with variable constraint value; $x_2 - x_1 = a$ and $x_{12} - x_{11} = a$; $x_3 - x_1 = b$ and $x_{13} - x_{11} = b$; $x_4 - x_1 = c$ and $x_{14} - x_{11} = c$. In the second step we note that there are six components $C_1 = \{x_1, x_2, x_3, x_4\}$, $C_2 = \{x_5\}$, $C_3 = \{x_6\}$, $C_4 = \{x_7\}$, $C_5 = \{x_8\}$, $C_6 = \{x_9, x_{10}, x_{11}, x_{12}, x_{13}, x_{14}\}$, and that no component contains more than one axis of symmetry.

In the third step we analyze the auxiliary problem involving the variables a,b,c. Vertices $x_1$ and $x_{11}$ are labeled 0, vertices $x_2$ and $x_{12}$ are labeled a, vertices $x_3$ and $x_{13}$ are labeled b, while vertices $x_4$ and $x_{14}$ are labeled c. The maximum weight path from 0 to a in either component (in this case from $x_1$ to $x_2$) leads to the constraint $a - 0 \geq 3$, the maximum weight path from a to b leads to the constraint $b - a \geq 2$, the maximum weight path from b to a leads to the constraint $a - b \geq -2$, etc. Thus we obtain the variable constraint graph in figure 9 below.

**Figure 9**

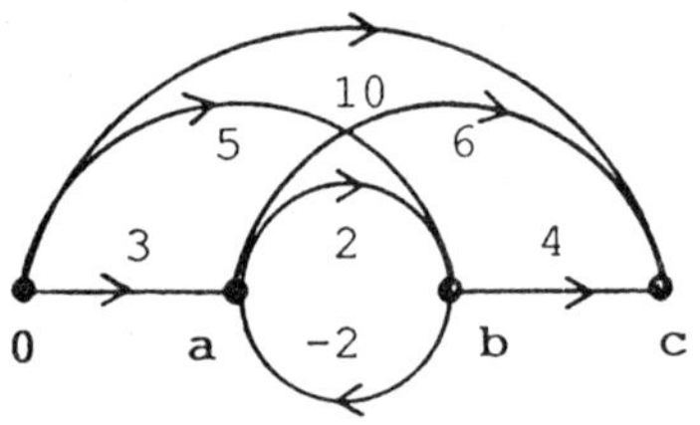


We can use the basic algorithm to obtain the minimum values a=3, b=5, and c=10. In step 4 we add the new constraints $x_2 - x_1 = 3$ and $x_{12} - x_{11} = 3$, $x_3 - x_1 = 5$ and $x_{13} - x_{11} = 5$, $x_4 - x_1 = 10$ and $x_{14} - x_{11} = 10$ to replace the original symmetry constraints with variable constraint values. Since the design rules are now expressed using constraints with numerical constraint values, the basic algorithm can now be applied to obtain the final values below.

$x_1 = 0$, $x_2 = 3$, $x_3 = 5$, $x_4 = 10$, $x_5 = 12$, $x_6 = 14$, $x_7 = 15$, $x_8 = 13$, $x_9 = 17$, $x_{10} = 19$, $x_{11} = 23$, $x_{12} = 26$, $x_{13} = 28$, $x_{14} = 33$.

On the other hand consider the same procedure when an axis of 3-variable symmetry is added. Let $x_{11}$ be an axis of symmetry with $(x_{12} - x_{11}) = (x_{11} - x_{10})$, $(x_{13} - x_{11}) = (x_{11} - x_9)$, and $(x_{14} - x_{11}) = (x_{11} - x_6)$. In the first step we add the following equalities with variable constraint value; $x_{12} - x_{11} = a$ and $x_{11} - x_{10} = a$; $x_{13} - x_{11} = b$ and $x_{11} - x_9 = b$; $x_{14} - x_{11} = c$ and $x_{11} - x_6 = c$. In the second step we find that there are four strongly connected components $C_1 = \{x_1\}$, $C_2 = \{x_2, x_3\}$, $C_3 = \{x_4\}$, and $C_4 = \{x_5, x_6, x_7, x_8, x_9, x_{10}, x_{11}, x_{12}, x_{13}, x_{14}\}$. Since there is only one axis of symmetry we can obviously apply the second generation algorithm. In the third step we analyze the auxiliary problem involving the variables a,b,c. Vertex $x_{11}$ which is the axis of symmetry is assigned a label of 0, vertex 12 is assigned label a, vertex 10 is assigned label -a, and so on. We compare the maximum weight paths from -a to 0 and from 0 to a to obtain the constraint $a - 0 \geq 2$, from -a to c and from -c to a to obtain the constraint $a + c \geq 14$, from a to b and -b to -a to obtain the constraint $b - a \geq 2$, from -a to a to obtain the constraint $a + a \geq 5$, etc. These constraints are represented on the constraint graph in figure 10 with vertices -c,-b,-a,0,a,b,c. In addition, there is a maximum weight path of weight 5 from -a to a, of weight 10 from -b to b, and of weight 20 from -c to c, so we add the constraints $a \geq 3$, $b \geq 5$, and $c \geq 10$.

**Figure 10**

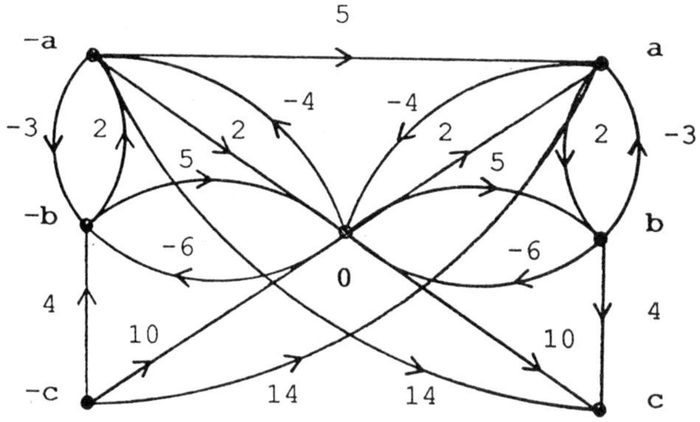


At this point we apply the basic algorithm to obtain the minimum values a=3, b=5, and c=10, so we set c=10 and add this constraint to the graph in figure 11. Then we apply the basic algorithm again to obtain minimum values a=4 and b=6, so we set b=6 and add this constraint to the graph in figure 10. We apply the basic algorithm a final time to obtain a=4. After substituting these numerical values for the variable constraint values we apply the basic algorithm to obtain the following x-values.

$x_1 = 0$, $x_2 = 3$, $x_3 = 5$, $x_4 = 8$, $x_5 = 10$, $x_6 = 12$, $x_7 = 13$, $x_8 = 11$, $x_9 = 16$, $x_{10} = 18$, $x_{11} = 22$, $x_{12} = 26$, $x_{13} = 28$, $x_{14} = 32$.

**45 Degree Line Constraints**

It is also desirable to allow 45 and -45 degree lines in an integrated circuit design. We need one 4-variable constraint to define the slope of such a line ($x_7 - x_6 = y_8 - y_4$). Notice that in this case the x-values and the y-values are no longer independent. When 45 degree lines are present in the design it is also convenient to allow constraints on the distance between a line and a corner point. If corner $(x_8, y_7)$ is required to be at least 2 units to the left of the 45 degree line and corner $(x_9, y_5)$ is required to be 4 units to the right of the 45 degree line in figure 11 then we obtain the constraints $(y_7 - y_4) - (x_8 - x_6) \geq \lceil 2\sqrt{2} \rceil = 3$, and $(x_9 - x_6) - (y_5 - y_4) \geq \lceil 4\sqrt{2} \rceil = 6$.

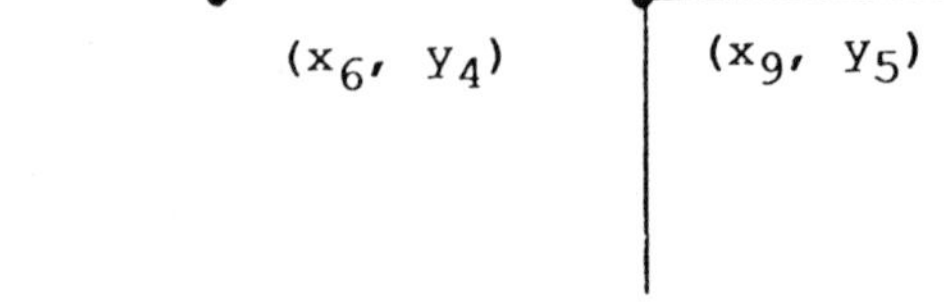

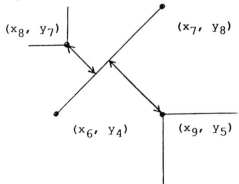

**Figure 11**

We can again replace the 45 degree line constraints by 2-variable constraints with variable constraint values. In this case the points $x_6$ and $y_4$, one in the x-constraint graph and one in the y-constraint graph, will also be referred to as axes. The second generation algorithm can also be applied to designs with a single 45 or -45 degree axis in a strongly connected component. Suppose that the following y-constraints are added to the list in figure 4 in addition to the 45 degree line constraints described above.

$y_1$; $y_2 - y_1 \geq 2$

$y_2$; $y_3 - y_2 \geq 2$, $y_4 - y_2 \geq 4$

$y_3$; $y_5 - y_3 \geq 4$

$y_4$; $y_5 - y_4 \geq 1$, $y_7 - y_4 \geq 3$

$y_5$; $y_6 - y_5 \geq 1$, $y_8 - y_5 \geq 3$

$y_6$; $y_7 - y_6 \geq 2$, $y_9 - y_6 \geq 5$

$y_7$; $y_8 - y_7 \geq 1$, $y_{10} - y_7 \geq 5$

$y_8$; $y_9 - y_8 \geq 3$

$y_9$; $y_{10} - y_9 \geq 2$

$y_{10}$; $y_{11} - y_{10} \geq 4$

$y_{11}$; empty

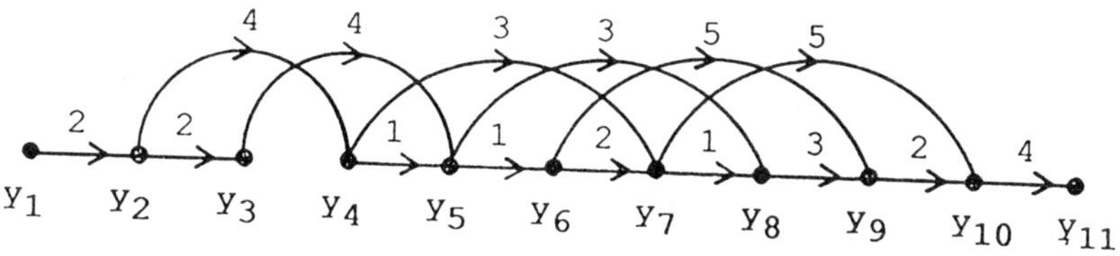


In this case the auxiliary problem for one component in the x-constraint graph and one component in the y-constraint graph must be considered simultaneously. The constraints $x_7 - x_6 = a$ and $y_8 - y_4 = a$ can be used to replace the constraint that defines the slope of the 45 degree line. Each point to line constraint also introduces two equalities with variable constraint values, $x_9 - x_6 = b$, $y_5 - y_4 = c$, and $y_7 - y_4 = d$, $x_8 - x_6 = e$, respectively. In the x-component vertex $x_6$ is labeled 0, vertex $x_7$ is labeled a, vertex $x_9$ is labeled b, and vertex $x_8$ is labeled e. In the y-component vertex $y_4$ is labeled 0, vertex $y_8$ is labeled a, vertex $y_5$ is labeled c, and vertex $y_7$ is labeled d. The variable constraint graph is constructed by finding the maximum weight paths between every pair of vertices in the corresponding x and y components. In addition the special constraints $b - c \geq 6$ and $d - e \geq 3$ are added to ensure the required point to line separations. We obtain the following variable constraint graph for variable weight edges a,b,c,d,e.

**Figure 13**

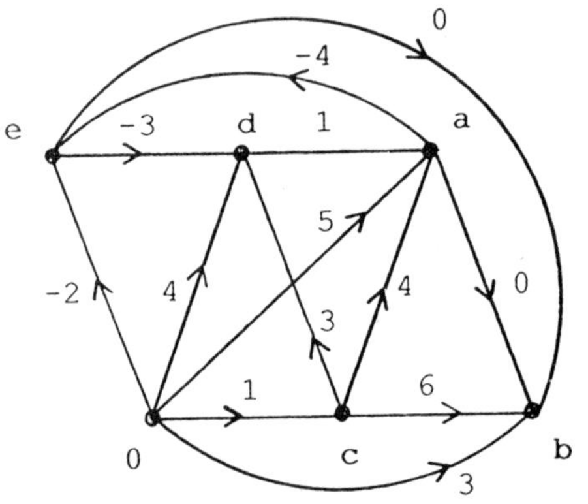


The auxiliary problem for a component containing a 45 degree axis will be similar to that for a component containing an axis of 4-variable symmetry. The minimum values a=5, b=7, c=1, d=4, e=1, can be obtained using the basic maximum path algorithm. Once these numerical values for the variable constraint values are determined, the x-values and the y-values can again be determined independently using the basic algorithm. To the x-constraints we first add $x_7 - x_6 = 5$, $x_8 - x_6 = 1$, $x_9 - x_6 = 7$, and to the y-constraints we add $y_8 - y_4 = 5$, $y_7 - y_4 = 4$, $y_5 - y_4 = 1$, to replace the 45 degree line constraints with variable constraint values. We then apply the basic algorithm to obtain the x-values and the y-values independently. The final values are listed below.

$x_1 = 0$, $x_2 = 3$, $x_3 = 5$, $x_4 = 8$, $x_5 = 10$, $x_6 = 12$, $x_7 = 17$, $x_8 = 13$, $x_9 = 19$, $x_{10} = 21$, $x_{11} = 23$, $x_{12} = 26$, $x_{13} = 28$, $x_{14} = 35$.

$y_1 = 0$, $y_2 = 2$, $y_3 = 4$, $y_4 = 7$, $y_5 = 8$, $y_6 = 9$, $y_7 = 11$, $y_8 = 12$, $y_9 = 15$, $y_{10} = 17$, $y_{11} = 21$.

**References**

1. B. Aspvall and Y. Shiloach, A Polynomial Time Algorithm for Solving Systems of Linear Inequalities with Two Variables per Inequality, SIAM J. Comput., Vol. 9(1980)pgs. 827-845.

2. P. Cook, Constraint Solver for Generalized IC Layout, IBM J. Res. Develop., Vol. 28(1984)pgs. 581-589.

3. S. Even, Graph Algorithms, Comp. Sci. Press, Rockville, 1979.

4. L. Liao, and C. Wong, An Algorithm to Compact a VLSI Symbolic Layout with Mixed Constraints, Proceedings of the 20th Design Automation Conference, June 1983, pgs. 107-112.

5. K. Mehlhorn, Data Structures and Algorithms, Vol. 2, Graph Algorithms and NP-Completeness, Springer-Verlag, New York, 1984.

6. T. Ohtsuki, Ed. Layout Design and Verification, Chap. 6, Layout Compaction, North Holland, Amsterdam, 1986.

7. R. Shostak, Deciding Linear Inequalities by Computing Loop Residues, Journal ACM, Vol. 28(1981)pgs. 769-779.

Contemporary Mathematics
Volume 89, 1989

# Parallelism, Preprocessing, and Reachability

Philip N. Klein

**Abstract**

The problem of reachability in a directed graph has resisted attempts at efficient parallelization. Only for fairly dense graphs can we efficiently achieve significant parallel speed-ups, using known methods. I describe a technique allowing significant speed-up even for moderately sparse graphs, following a preprocessing step in which a representation of the graph is created.

## 1 Introduction

In this paper we demonstrate the usefulness of preprocessing a graph in order to enable reachability queries to be processed quickly in parallel.

Parallel processing offers potentially vast improvements in computational performance. However, in applying parallel processing to specific problems, we encounter a serious obstacle: for some problems, even the best of known algorithms permit only a very small speed-up in relation to the number of processors used. That is, while potentially we can reduce the time for solving a problem by a factor of $p$ when $p$ processors are used, in fact the speed-up factor may be closer to $\sqrt{p}$ or less. Indeed, for some problems, to achieve even a moderate speed-up seems to require an enormous number of processors.

One fundamental such problem is the following *directed reachability problem*: Let $G$ be a directed graph with $n$ nodes and $m$ edges. For a given set $S$ of nodes of $G$, find

1980 *Mathematics Subject Classification* (1985 *Revision*), 05C20, 06A10, 68Q25

Author's research supported by an ONR Graduate Fellowship, by AT&T Bell Laboratories, and by Air Force Contract AF0SR-86-0078


0271-4132/89 $1.00 + $.25 per page

the set $T$ of nodes reachable in $G$ from nodes of $S$. We call $S$ the set of "sources," and $T$ the set of "targets". While this problem has a simple linear-time sequential solution (time $O(m+n)$), it has so far proved difficult to parallelize efficiently.

This problem or closely related problems arise in connection with diverse other problems. In artificial intelligence: a basic operation arising in semantic network manipulation is identifying the lowest common ancestors of some nodes. In combinatorial optimization: finding an augmenting path while solving a matching or flow problem is essentially finding a path in a directed graph. In databases: handling recursive queries can be reduced to finding paths between nodes in a directed graph.

Two approaches to the directed reachability problem are known, parallel transitive closure and parallel breadth-first search. There is a parallel algorithm for computing the transitive closure of $G$ in $O(\log^2 n)$ time. However, this algorithm seems to require about $n^3$ processors for all practical purposes; while asymptotically better algorithms exist, the associated constants are huge. Moreover, even supposing the transitive closure was already computed, using it to solve the directed reachability problem requires $\Omega(n^2)$ operations. Hence this approach is not efficient in comparison to the $O(m+n)$ time sequential solution: very many processors are needed in relation to the speed-up achieved.

On the other hand, a parallel version of breadth-first traversal of the graph $G$, starting at the nodes of $S$, can, over a sequence of stages, identify all the reachable nodes. Initially, we mark the nodes in $S$. In each stage, we consider the set of edges leaving newly marked nodes, and mark the nodes they enter, until the stage at which no additional nodes are marked. The number of stages needed is $\Omega(n)$ in the worst case, so the use of parallelism cannot guarantee a speed-up of more than $O((n+m)/n)$. Thus for moderately sparse graphs, this method fails to be useful in the worst case.

We propose a two-part solution. First, the graph $G$ is preprocessed sequentially, and a representation of $G$ is created. The representation depends on the number $p$ of processors we intend to use. Our technique works only for values of $p \leq (m/n)\sqrt{m}$. Once the representation exists, we can use $p$ processors to answer queries of the form "given sources $S$, find targets $T$" in $O((n+m)/p)$ parallel time. The approach incorporates a trade-off between speed of processing a query, and compactness of the representation of $G$. The storage required by the representation is $O(n+m+n^2p/m)$, which is optimal when $p \leq m^2/n^2$. The method for handling a query is simple enough to be potentially quite practical.

In related work, Gambosi, Nešetřil, and Talamo [4] describe techniques for preprocessing a directed acyclic graph in order to facilitate sequentially searching for a path between two given nodes.

# 2 Preprocessing

We assume henceforth for simplicity that the the number $m$ of arcs of $G$ is at least the number $n$ of nodes, minus one. Otherwise, the underlying graph is disconnected, and we may independently consider each connected component.

## 2.1 Eliminating cycles

First we reduce the problem to the case in which $G$ is acyclic. Given any directed graph $G$, we find the strongly connected components of $G$, and obtain a graph $\hat{G}$ from $G$ by contracting each strongly connected component to a node. The strongly connected components of $G$ can be computed in $O(n+m)$ sequential time using an algorithm of Tarjan [8].

For any node $v$ of $G$, let $\hat{v}$ denote the node of $\hat{G}$ corresponding to the strongly connected component of $G$ containing $v$. Now, given a set $S$ of sources in $G$, let $\hat{S} = \{\hat{v} : v \in S\}$. If $\hat{T}$ is the corresponding set of targets in $\hat{G}$, let $T = \{v : \hat{v} \in \hat{T}\}$. The correctness of this reduction follows immediately from the definition of strongly connected components.

We therefore assume henceforth that $G$ is an directed acyclic graph, or *dag*. Hence $G$ defines a partial order on its nodes, namely $v < w$ if $w$ is reachable from $v$. Let $\mathcal{P}(G)$ denote the partially ordered set $(V(G), <)$ that the dag $G$ thus defines on its nodes.

## 2.2 The antichain-chain decomposition

We review some terminology of partially ordered sets, or *posets*. For a poset $\mathcal{P} = (V, <)$, the poset $\mathcal{P}' = (V', <')$ is a *subposet* of $\mathcal{P}$ if $V' \subseteq V$ and for every pair of nodes $v, w \in V'$, we have $v <' w$ iff $v < w$. A *chain* is a set of nodes every two of which are related by $<$. The nodes in a chain can be totally ordered: $v_1 < v_2 < \cdots < v_k$. An *antichain* is a set of nodes no two of which are related. A *chain cover* is a partition of the nodes into chains; an *antichain cover* is a partition of the nodes into antichains. Clearly, the size of any chain cover is at least the size of any antichain, because each chain can cover at most one element of the antichain. Similarly, the size of any antichain cover is at least the size of any chain. In fact,

**Dilworth's Theorem[2]:** *In any partially ordered set (poset), the size of the smallest chain cover equals the size of the largest antichain.*

The following theorem is well known. It appears, for example, as Prop. 8.15 of p. 398 of [1].

**Dual of Dilworth's Theorem:** *In any poset $\mathcal{P}$, the size of the smallest antichain cover equals the size of the largest chain.*

**PROOF:** For each node $v$, let the *rank* $r(v)$ of $v$ be the number of nodes in the longest chain ending at $v$. Let $k$ be the maximum rank of a node in $\mathcal{P}$. Then for each $1 \leq i \leq k$, the set of nodes of rank $i$ is an antichain, so we have an antichain cover of size $k$. Since the size of any antichain cover is at least the size of any chain, it follows that we have found a minimum-size antichain cover and a maximum-size chain. □

An inductive characterization of rank for a poset $\mathcal{P}(G)$ is as follows: when all nodes of rank 1 through $j$ are deleted, a remaining node is rank $j+1$ if and only if it now has no incoming arcs. This characterization suggests a sequential algorithm that, given a dag $G$, finds the ranks of $\mathcal{P}(G)$ in $O(m+n)$ time.

The algorithm we give is slighly more general in that it handles any subposet $\mathcal{P}'$ of $\mathcal{P}(G)$. We represent the subposet by assigning a flag *deleted*$[v]$ to each node $v$. The value of *deleted*$[v]$ is *false* for each node $v$ in the subposet, and *true* otherwise. For

**FIND-RANKS**

R1 Let $L_0 := \{\perp\}$, and let $r := 0$. For each node $v$, assign to $d[v]$ the indegree of $v$.

R2 Copy $L := L_r$.

R3 While $L$ is not empty, remove a node $v$ from $L$, and consider each arc $(v, w)$ in turn. Reduce $d[w]$ by 1. If $d[w]$ thereby becomes zero, then ...

R4 Set $f[w] := \begin{cases} v & \text{if } \mathit{deleted}[v] = \mathit{false} \\ f[v] & \text{if } \mathit{deleted}[v] = \mathit{true} \end{cases}$

R5 If $\mathit{deleted}[w] = \mathit{false}$, put $w$ on the list $L_{r+1}$ of rank $r+1$ nodes. Otherwise, put $w$ on the list $L$.

R6 If $L_{r+1}$ is not empty, set $r := r + 1$, and go to step 2. Otherwise, $r$ is the maximum rank.

Figure 1: Algorithm for finding the ranks of nodes of a subposet $\mathcal{P}'$ of $\mathcal{P}(G)$

notational convenience, add the node $\perp$ and the arcs $(\perp, v)$ for each $v$ in $G$, and set $\mathit{deleted}[\perp] = \mathit{false}$. The algorithm is given in Figure 1. It places all nodes of rank $r$ in the list $L_r$, and computes the maximum rank. The algorithm simultaneously constructs a table $f[\cdot]$ such that, for each node $v$ of rank more than 1, $f[v]$ is the penultimate node of a longest chain ending at $v$.

To find a maximum-size chain, let $k$ be the maximum rank, and let $\hat{v}$ be a node on the list $L_k$. Then $\hat{v}, f[\hat{v}], f[f[\hat{v}]], \ldots, f^{k-1}[\hat{v}]$ traces backwards along a $k$-node path from a rank 1 node to $\hat{v}$.

The following corollary, which follows from either Dilworth's Theorem or its dual, is the basis for our representation of dags:

**COROLLARY 2.1** *For any $1 \leq s \leq n$, an $n$-element poset can be decomposed into at most $s$ chains, each of size $\lceil n/s \rceil$, and at most $\lceil n/s \rceil - 1$ antichains.*

**PROOF:** If the poset has a chain of size $\lceil n/s \rceil$, remove it. Iterate this step until all remaining chains are of size less than $\lceil n/s \rceil$, so the poset remaining has an antichain cover of size less than $\lceil n/s \rceil$. Since each iteration removes $\lceil n/s \rceil$ elements, at most $s$ iterations are needed. □

**Definition:** A decomposition of an $n$-element poset $\mathcal{P}$ into sets $\mathcal{A}, C_1, \ldots, C_s$ is called an *antichain-chain decomposition* if $C_1, \ldots, C_s$ are chains of size at most $\lceil n/s \rceil$, and $\mathcal{A}$ is the union of at most $\lceil n/s \rceil - 1$ antichains.

To preprocess a dag $G$, we let $s = np/m$, where $p$ is the desired number of processors, and we find an antichain-chain decomposition $\mathcal{A}, C_1, \ldots, C_s$ of the poset $\mathcal{P}(G)$ defined by $G$. To find such a decomposition, we implement the proof of Corollary 2.1, using the procedure described above. Each iteration consists of finding the ranks of the poset corresponding to the current graph (and the table $f[\cdot]$), and then identifying a chain of size $\lceil n/s \rceil$ and deleting the nodes in this chain. After at most $s$ iterations, there is no chain of this size. Thus the decomposition can be found in $O(sm)$ sequential time. In the next subsection, we discuss further processing of the subposet $\mathcal{P}'$ of $\mathcal{P}(G)$ induced on the node-set $C_1 \cup \cdots \cup C_s$.

## 2.3 Processing of the chains: The earliest-entry table

Next we describe a table that facilitates reachability queries for a dag whose poset has a small chain cover.

**Definition:** A *top element* of a poset $\mathcal{P}$ is an element $\top$ that is bigger than all other elements of the poset. For any poset $\mathcal{P}$ with a top element $\top$, for a chain $C$ of $\mathcal{P}$, and for any node $u$ of $\mathcal{P}$, we define *$u$'s earliest entry into $C$* to be the minimum node $x$ in $C \cup \{\top\}$ reachable from $u$. The minimum is well-defined because $C \cup \{\top\}$ is a chain.

Given that $u$'s earliest entry into $C$ is $x$, we can determine exactly which nodes $v$ in the same chain $C$ are reachable from $x$; namely, $v$ is reachable if and only if $v \geq x$.

Earlier, we derived a poset $\mathcal{P}(G)$ from the dag $G$, and decomposed $\mathcal{P}(G)$ into some antichains and $s$ chains $C_1, \ldots, C_s$ of size at most $\lceil n/s \rceil$. Let $\mathcal{P}'$ be the subposet of $\mathcal{P}(G)$

induced by the union of the $s$ chains. That is, the elements of $\mathcal{P}$ are the elements of the $s$ chains, and the order relation of $\mathcal{P}'$ is simply the order relation of $\mathcal{P}(G)$ restricted to the elements of $\mathcal{P}'$. For notational convenience, add a top element $\top$ to $\mathcal{P}'$. Let $n' = 1 + \sum_i |C_i|$ be the number of elements of $\mathcal{P}'$.

To preprocess $\mathcal{P}'$, we compute an $n' \times s$ table $R[\cdot,\cdot]$ that is a generalization of the transitive closure of $G'$. For each element $u$ of $\mathcal{P}'$, and each $i \in \{1,\ldots,s\}$, let $R[u,i]$ be $u$'s earliest entry into $C_i$. We call this table the *earliest-entry table*. It requires $O(n's)$ storage.

We also compute two $n'$-element tables: we let $chain[u]$ be the index $j$ of the chain $C_j$ in which $u$ appears, and we let $rank[u]$ be the rank of $u$ in that chain. Let $rank[\top] = n'$.

The earliest-entry table $R[\cdot,\cdot]$, chain table $chain[\cdot]$, and rank table $rank[\cdot]$ can be used to determine reachability between any two nodes $u$ and $w$ of $\mathcal{P}'$. Namely, $u < w$ if and only if $rank(x) \leq rank(w)$, where $x = R[u, chain(w)]$.

The above approach to representing reachability is derived directly from a theorem in the dimension theory of posets stating that a poset with largest antichain of size $\omega$ has dimension at most $\omega$. Dimension theory is concerned with the compactness of representation of a poset as the intersection of total orders. See [7] for a survey. Also, Jagadish [6] independently considered the use of a Dilworth decomposition for more efficiently computing and more compactly representing the transitive closure of a directed graph, specifically in application to handling recursive database queries.

The tables $chain[\cdot]$ and $rank[\cdot]$ can easily be computed from the chain decomposition of $\mathcal{P}'$. For each node $w$ of $G$ that does not appear in the chain decomposition, let $chain[w] = 0$. We next observe that the earliest-entry table can be computed in $O(sm)$ sequential time. In particular, for each chain $C_j$, we show how in $O(m)$ time one can compute all the entries $R[v,i]$ of the table for $v \in C_j, 1 \leq i \leq s$. To do this, we avoid explicitly computing the poset $\mathcal{P}'$, and instead work directly from the dag $G$.

Suppose the chain $C_j$ consisted of a single node $u$. We could use, say, directed depth-first traversal of the dag $G$ rooted at $u$ to identify all the nodes $v$ of $G$ reachable

PROCESS-CHAIN

Initially, unmark all nodes, and set $R[v_{k+1}, i] = \top$ for $i = 1, \ldots, s$.

For $\ell = k, k-1, \ldots, 1$, do

  Initialize $\rho[i] := R[v_{\ell+1}, i]$ for $i = 1, \ldots, s$.

  Call the recursive procedure $VISIT(v_\ell)$.

  Copy $R[v_\ell, i] := \rho[i]$ for $i = 1, \ldots, s$.

Figure 2: The algorithm PROCESS-CHAIN for filling out the earliest-entry table.

from $u$. We maintain an $s$-element table $\rho[\cdot]$, initialized to $\top$. Whenever we visit a node $v$ such that $chain[v] = i \neq 0$, we compare $rank[v]$ to $rank[\rho[i]]$; if it is less, we set $\rho[i] := v$. When the traversal is finished, $\rho[\cdot]$ will be the minimum node of $C_i \cup \{\top\}$ reachable from $u$.

To achieve $O(m)$ time, depth-first search marks each node as it visits it, and avoids visiting a node that has previously been marked. Thus it avoids redundant search. To process a chain $C_j = (v_1 < v_2 < \cdots < v_k)$ consisting of more than one node, we imitate this idea. We start with a depth-first traversal rooted at $v_k$, and visit all nodes reachable from $v_k$, maintaining our table $\rho[\cdot]$. Next, we continue by performing a depth-first traversal rooted at $v_{k-1}$ *without first removing the marks placed on nodes by the first depth-first traversal.* Thus in the second traversal we avoid visiting nodes already visited in the first. Since $v_k$ is reachable from $v_{k-1}$, every node visited in the first traversal is reachable from $v_{k-1}$. The information we need about these nodes is conveniently summarized in the table $\rho[\cdot]$, so there is no reason to visit them again. The process continues in this fashion, considering the nodes of the chain $C_j$ in reverse order.

We now give a more formal description. Let $C_j$ be the chain $v_1 < \ldots < v_k$. For notational convenience, let $v_{k+1}$ denote $\top$. The algorithm for processing the chain $C_j$ appears in Figure 2. The subroutine VISIT($v$) is shown in Figure 3.

*VISIT*($v$):

V1 Mark $v$ as having been visited.

V2 If $chain(v) = i \neq 0$ and $rank[v] < rank[\rho[i]]$, then set $\rho[i] := v$.

V3 For each outgoing arc $(v, w)$, if $w$ is not already marked, call VISIT($w$).

Figure 3: The recursive subroutine VISIT($v$) for visiting reachable nodes.

The time required by the above algorithm is dominated by the time for step 4. Use of the marks to truncate the depth-first search ensures that no arc is explored more than once during the execution of the above procedure. Hence the total time spent in step 4 is $O(m)$.

We now consider correctness of the procedure. For $i = k, k-1, \ldots, 1$, let $W_i$ be the set of nodes $w$ visited during iteration $i$ of step 4. (Let $W_{k+1} = \{\top\}$). It can be shown by a simple backwards induction on $\ell \leq k+1$ that the set of nodes reachable from $v_\ell$ is $W_\ell \cup W_{\ell+1} \cup \cdots \cup W_{k+1}$, and that for $i = 1, \ldots, s$, $R[v_\ell, i]$ is the minimum node $w$ in $C_i \cup \{\top\}$ reachable from $v_\ell$.

# 3 Answering a query

## 3.1 Using the earliest-entry table

In this section, we give a parallel method for answering a reachability query for a poset with a small chain cover, using the earliest-entry table defined in Subsection 2.3. In particular, we consider the poset $\mathcal{P}'$ with a chain cover consisting of $s$ chains $C_1, \ldots, C_s$ of size at most $\lceil n/s \rceil$.

Consider a single chain $C_1$, and a subset $S$ of the nodes of $C_1$. The nodes of $C_1$ reachable from nodes in $S$ are exactly the nodes $v \geq x$, where $x$ is the minimum node in $S$. Thus for purposes of reachability, we may as well replace the entire set $S$ with the single node $x$.

More generally, suppose $S$ is a subset of the nodes of chains $C_1, \ldots, C_s$. Let $x_i$ be the minimum node in $S \cap C_i$ for $i = 1, \ldots, s$. We may as well replace the set $S$ with the $s$-element set $\{x_1, \ldots, x_s\}$. Since each chain $C_i$ has size at most $\lceil n/s \rceil$, the node $x_i$ can easily be found in $O(n/s)$ time by scanning up the chain $C_i$ in order until a node of $S$ is encountered. Hence the set $\{x_1, \ldots, x_s\}$ can be found in $O(n/s)$ time using $p \geq s$ processors.

Now for each chain $C_j$, the minimum node $y_j$ reachable from any of the nodes $x_1, \ldots, x_s$ is just the minimum of the earliest entries of $x_1, \ldots, x_s$ into $C_j$. That is, using the earliest-entry table, we have $y_j = \min\{R[x_i, j] \ : \ i = 1, \ldots, s\}$. We can find the minimum in $O(s + n/s)$ time by marking all the earliest entries into $C_j$ (in $O(s)$ time) and then scanning up the chain $C_j$ in order until a marked node is encountered (taking $O(n/s)$ time). The first of these two steps, marking the earliest entries, can be easily parallelized; simply divide the set of $s^2$ pairs $(x_i, C_j)$ evenly among the $p$ processors, and let each processor be responsible for $s^2/p$ marking operations. Thus the set $\{y_1, \ldots, y_s\}$ can be found in $O(s^2/p + n/s)$ time using $p$ processors.

Finally, to find the set $T_j$ of all reachable nodes in the chain $C_j$, we let $T_j$ be the set of nodes in the chain that have rank at least that of $y_j$. These nodes can be marked by yet another scan up the chain (starting at $y_j$) taking $O(n/s)$ time. The set $T = T_1 \cup \cdots \cup T_s$ of all reachable nodes can be determined in $O(n/s)$ time using $p$ processors.

The total time for the above procedure is $O(s^2/p + n/s)$ using $p$ processors. We now consider correctness. Suppose $v$ is a node of $S$ and a node $w \in C_j$ is reachable from $v$. By definition of $x_i$, $x_i \leq v$. Since $v \leq w$, certainly $x_i \leq w$, so $x_i$'s earliest entry $R[x_i, j]$ into $C_j$ is $\leq w$. Hence the minimum earliest entry $y_j$ is $\leq w$, so $w \in T_j$. This argument shows that $T$ contains all nodes reachable from nodes in $S$.

## 3.2 Solving the directed reachability problem on a union of antichains

In Subsection 2.2 we showed that the nodes of a dag $G$ could be decomposed into a set $\mathcal{A}$ and a poset $\mathcal{P}'$ such that $\mathcal{A}$ consists of at most $n/s$ antichains, and the poset $\mathcal{P}'$

consists of at most $s$ chains. In Subsection 3.1, we described a parallel solution to the directed reachability problem for $\mathcal{P}'$. In this section, we consider the same problem as applied to the subgraph $\widehat{G}$ of $G$ induced by $\mathcal{A}$.

Observe that a directed path cannot contain two nodes in a single antichain. Since $\widehat{G}$ consists of at most $n/s$ antichains, any directed path in $G$ contains at most $n/s$ nodes. Recall the method of "parallel breadth-first search" sketched in the introduction: start by marking the nodes in $S$; in each stage, consider the arcs leaving newly marked nodes, and mark the nodes they enter. If there is a path in $\widehat{G}$ from $x$ to $y$, the path contains at most $n/s$ nodes, so $n/s$ stages of parallel breadth-first search are sufficient to identify all reachable nodes.

Let $m_i$ be the number of arcs considered in stage $i$, for $i = 1, \ldots, n/s$. Each arc is considered at most once (immediately after the node it leaves has been marked), so $\sum_i m_i$ is no more than $m$, the number of arcs in $G$. The number of operations needed to carry out stage $i$ is $O(m_i)$. With $p$ processors available, therefore, stage $i$ can be carried out in $O(\lceil m_i/p \rceil)$ time. Summing over all stages yields $O(m/p + n/s)$ parallel time to solve the directed reachability problem on $\widehat{G}$.

One might object that the above analysis has omitted the time to assign arcs to processors. Using the notion of *ranks* outlined in Subsection 2.2, we can choose such an assignment during preprocessing that will work regardless of the choice of $S$. The key to the the assignment is the following property of partitioning into ranks: if a rank $i$ node is reachable from another node, it is reachable via a path of length at most $i$.

Let $M_i$ be the set of arcs leaving nodes of rank $i$, for $i = 1, \ldots, n/s - 1$. During preprocessing, we evenly divide up the arcs of $M_i$ among the $p$ processors, for each $i$. Now to answer a query, we first mark the nodes of $S$, and then carry out a modified version of breadth-first search in which, in stage $i$, we consider the set $M_i$ of arcs. Each processor considers one by one the arcs of $M_i$ assigned to it; for each such arc, the processor marks the arc's head if the arc's tail is already marked. It is easy to verify that after stage $i$, the set of marked nodes consists of nodes of $S$, together with all nodes of rank $\leq i$ reachable from nodes of $S$.

## 3.3 Combining the techniques

In this subsection, we observe that the techniques used in Subsections 3.1 and 3.2 may be combined to yield a solution for the dag $G$. Suppose we have found an antichain-chain decomposition $A, \mathcal{C}_1, \ldots, \mathcal{C}_s$ of $G$ as described in Subsection 2.2, we have preprocessed the poset $\mathcal{P}'$ induced by the chains, forming the earliest-entry table, and we have preprocessed the graph $\widehat{G}$ induced on the antichains $\mathcal{A}$, choosing an assignment of arcs to processors.

Suppose we are given a set $S$ of sources to process. We write $S = S^a \cup S^c$, where $S^a$ consists of "antichain" nodes, *i.e.* nodes in $\mathcal{A}$, and $S^c$ consists of "chain nodes," nodes appearing in some $\mathcal{C}_i$. The algorithm consists of an "antichain" step, a "chain" step, and a final "antichain" step, with transitional steps in between:

**Step 1 (Antichain step)** Apply the technique of Subsection 3.2 to find the set $T_1$ of nodes reachable from $S^a$ via paths entirely in $\widehat{G}$.

**Step 2 (Transitional step)** Consider the set of arcs from antichain nodes to chain nodes, marking each chain node that is the head of an arc whose tail is an antichain node in $T_1$. Let $S'$ be the set of chain nodes thus marked.

**Step 3 (Chain step)** Apply the technique of Subsection 3.1 to find the set $T_2$ of nodes reachable from $S^c \cup S'$ (i.e. $T_2 = \{v \in \mathcal{P}' \ : \ u < v \text{ for some } u \in S^c \cup S'\}$).

**Step 4 (Transitional step)** Consider the set of arcs from chain nodes to antichain nodes, marking each antichain node that is the head of an arc whose tail is a chain node in $T_2$. Let $S''$ be the set of antichain nodes thus marked.

**Step 5 (Antichain step)** Apply the technique of Subsection 3.2 to $S''$, yielding a set $T_3$ of antichain nodes. We return the set $T = T_1 \cup T_2 \cup T_3$.

We now consider the correctness of the above procedure for computing the set $T$ of targets from the given set $S$ of sources. It is easy to see that every node in $T$ is in fact reachable from a node in $S$. To see that every reachable node $w$ is in $T$, let

$v_1 v_2 \ldots v_k = w$ be any directed path in $G$, where $v_1$ belongs to $S$. If this path consists entirely of nodes of $\mathcal{A}$, then $w$ is in $T_1$. Suppose therefore that the path contains at least one node not in $\mathcal{A}$; let $v_i$ be the first such node, and let $v_j$ be the last. If $i = 1$ then $v_i \in S^c$. If $i > 1$ then $v_{i-1} \in T_1$ after step 1, so $v_i \in S'$ after step 2. In either case $v_i \in S^c \cup S'$. Since $v_i < v_j$ in the poset $\mathcal{P}'$ induced on the chain nodes, we have that $v_j \in T_2$ after step 3. If $v_j = w$, we are done; otherwise, step 4 ensures that $v_{j+1} \in S''$, and hence step 5 ensures that $w \in T_3$.

Next we consider the complexity of the procedure. Steps 1 and 5 take $O(m/p+n/s)$ time using $p$ processors. Each of steps 2 and 4 is essentially a single stage of parallel breadth-first search, and hence takes $O(m/p)$ time. Step 3 takes $O(n/s + s^2/p)$ time using $p \geq s$ processors. Hence the procedure takes $O(m/p + n/s + s^2/p)$ time using $p$ processors.

In order that the procedure take time $O(m/p)$, we need to choose $s$ so that $n/s \leq m/p$ and $s^2 \leq m$; that is, $np/m \leq s \leq \sqrt{m}$. We can achieve this when $p \leq m^{1.5}/n$ by choosing $s = np/m$. With this value of $s$, the storage needed for the results of preprocessing is $O(m + ns) = O(m + n^2 p/m)$, which is optimal when $p \leq (m/n)^2$.

We evidently have a greater range of choice for our speed-up $p$ when the graph is denser. This is not surprising; both of the algorithms described in the introduction, computing transitive closure and parallel breadth-first search, make better use of parallelism when the input graph is very dense. In contrast to these two algorithms, however, our algorithm can achieve optimal speed-up of $O(\sqrt{n})$ no matter how sparse the graphs (although for very sparse graphs, the storage is not optimal). Moreover, for even moderately sparse graphs, say $m \approx n^{1.3}$, we can achieve optimal speed-up by a factor of $\approx n^{.95}$—and optimal storage as well if the speed-up is no more than $n^{.6}$.

# 4 Concluding remarks

We have presented a method for representing a directed graph in a way that permits fast parallel answers to reachability queries. Preprocessing is proposed as a way of coping with our current inability to efficiently solve the general, unpreprocessed problem in parallel.

The method can be extended to handle reachability in a graph slightly different from the original, preprocessed graph $G$. That is, given a set $S$ of sources, a set $E$ of edges to be added, and a set $F$ of edges to be removed, we can in $O(m/p + |E| + |F|)$ parallel time find the set $T$ of nodes reachable from nodes of $S$ in the graph $G \cup E - F$.

This extension to handling a slighly modified graph suggests the following open question: can updates to the graph representation be efficiently carried out in parallel? That is, can the representation be dynamically modified?

# Acknowledgements

Thanks to David Shmoys for advising me during this research. Thanks to Ramesh Patil for suggesting the problem. Thanks to Mike Saks for noting that Corollary 2.1, which was originally proved using Dilworth's Theorem, could be proved using the dual of Dilworth's Theorem, thereby simplifying the preprocessing. Thanks to Bruce Maggs for a helpful discussion.

## References

[1] , M. Aigner, *Combinatorial Theory*, Springer-Verlag, New York (1979)

[2] R. P. Dilworth, "A decomposition theorem for partially ordered sets, " *Ann. Math.*, *51* (1950), pp. 161-166

[3] L. R. Ford, Jr., and D. R. Fulkerson, *Flows in Networks*, Princeton University Press, Princeton, NJ (1962)

[4] G. Gambosi, J. Nešetřil, and M. Talamo, "Posets, Boolean representations, and quick path searching," Proc. ICALP 87, Thomas Ottman, ed., published as *Lecture Notes in Computer Science 267*, Springer-Verlag, New York, pp. 404-424.

[5] Hiraguchi, "On the dimension of orders," Sci. Rep. Kanazawa Univ. 4 (1955), pp. 1-20

[6] H. V. Jagadish, "A compressed transitive closure technique for efficient fixed-point query processing," manuscript (1987), AT&T Bell Labs, Murray Hill, New Jersey

[7] D. Kelly and W. T. Trotter, "Dimension theory for posets," in *Ordered Sets: Proceedings of the NATO Advanced Study Institute held at Banff, Canada, 1981*, I. Rival, ed., D. Reidel Publishing Co., Boston (1982), pp. 171-211

[8] R. Tarjan, "Depth-first search and linear graph algorithms," *SIAM J. Computing* 1:2 (1972), pp. 146-160

LABORATORY FOR COMPUTER SCIENCE
MASSACHUSETTS INSTITUTE OF TECHNOLOGY
CAMBRIDGE, MASSACHUSETTS 02139

Contemporary Mathematics
Volume **89**, 1989

# A SUMMARY OF RESULTS ON PAIR-CONNECTED RELIABILITY

Peter J. Slater[1,2]

ABSTRACT. A general formula that encompasses the various reliability measures in the literature is presented. Each instance of the general formula can be interpreted as the expected value of a random variable. In particular, global connectivity and two-terminal connectivity are instances in which the random variable is an indicator variable.

The pair-connected measure of network reliability is presented as the expected value of a sum of indicator variables, and some known results for this measure are summarized.

I. INTRODUCTION. A widely used model for communications networks in which elements are subject to failure is that of a probabilistic graph G with vertex set V(G) and edge set E(G). For $|V(G)| = n$ and $|E(G)| = m$ such a graph G is referred to as an (n, m)-graph. Here it is assumed that vertices are fail-safe but that each edge $e \in E(G)$ is down (that is, in a failed state) independently with probability q where $0 < q < 1$. Let $p = 1 - q$ denote the probability that each edge is up. For $S \subseteq E(G)$, G is said to be in state S if the edges that are up are precisely those in S. Let $<S>$ denote the spanning subgraph of G with edge set S. The following is a general formula for reliability measures associated with such graphs.

$$(1) \qquad R(G,q,f) = \sum_{S \in \Omega} f(S) \bullet R(S)$$

In (1) $\Omega$ is the collection of all subsets of E (that is, the set of all possible states for the system). If $S \in \Omega$ then R(S) denotes the probability that G is in state S which, under the assumptions of independence and equal probability q of failure, means $R(S) = p^k q^{m-k}$ when $|S| = k$. Note that $(\Omega, R)$ is a probability space, that f is a random variable defined in this space, and that R(G,q,f) is the expected value of f.

Different choices for the function f provide a variety of reliability measures. If we let $f(S) = 1$ or $0$ if $<S>$ is connected or disconnected, respectively, then (1) is the formula for

1980 Mathematics Subject Classifications: 05C99, 06A10, 62N05, 68M20.

[1]Research supported in part by the U.S. Office of Naval Research Grant N00014-86-K-0745.

[2]The author thanks Clemson University for its support while he visited in the Fall, 1987.

global connectivity or all terminal reliability, the probability $R(G; q)$ that $G$ is connected. For two terminal reliability with designated vertices $s$ and $t$, if $f(S) = 1$ or $0$ if $s$ and $t$ are connected or disconnected in <S>, respectively, then (1) is the formula for $R_{s,t}(G; q)$, the probability that $s$ and $t$ are connected. For these instances the random variable $f$ is an indicator variable (taking only zero-one values).

Simply noting that one frequently encounters measures of unreliability rather than reliability, [6, 11, 19] are excellent surveys concerning network reliability, and a selection of articles from the network reliability literature is [5, 7, 8, 16, 17, 18, 21].

Previous work related to "pair-connected" reliability and the related concept of "cutting numbers" includes [8, 10, 12, 13, 14, 19]. The pair-connected reliability of $G$ is the expected number of pairs of connected vertices. Specifically, in (1) let $f(S) = PC(S)$ equal the number of pairs of vertices that are connected in <S>. Then the pair-connected reliability of $G$ is given as follows.

$$(2) \qquad PC(G; q) = \sum_{S \in \Omega} PC(S) \bullet R(S)$$

Here I summarize some of the results in [1, 2, 3, 4] concerning pair-connected reliability (which is called "resilience" in [9]).

2. PAIR-CONNECTED RELIABILITY. For the graph $H$ in Figure 1 three states corresponding to situations with three failed edges are $S1 = \{ab, af, bf, cd, ce, de\}$, $S2 = \{af, ab, bf, bc, cf, de\}$ and $S3 = \{bc, bf, cd, ce, de, ef\}$. Because $PC(S1) = 6$, $PC(S2) = 7$ and $PC(S3) = 10$, three of the terms in summation (2) are $6p^6q^3$, $7p^6q^3$ and $10p^6q^3$. In general, the pair-connected reliability polynomial can be written as

$$(3) \qquad PC(G; q) = B_1 p q^{m-1} + B_2 p^2 q^{m-2} + \ldots + B_{m-1} p^{m-1} q + B_m p^m$$

where each $B_i$ represents the total number of pairs of connected vertices taken over all subgraphs with exactly $i$ edges. For graph $H$ in Figure 1 we have $PC(H; q) = 9pq^8 + 92p^2q^7 + 411p^3q^6 + 1020p^4q^5 + 1463p^5q^3 + 1155p^6q^3 + 530p^7q^2 + 135p^8q + 15p^9$.

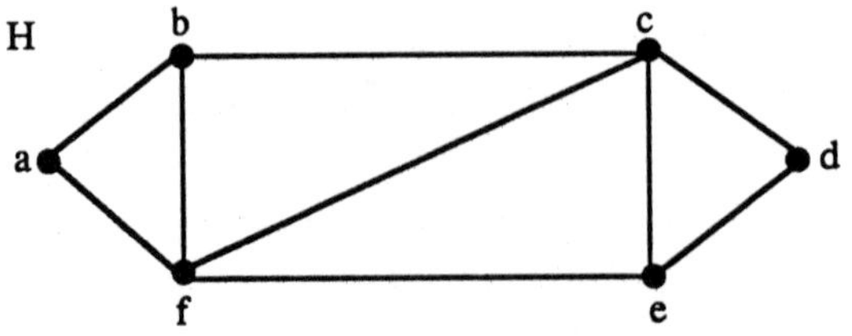


Figure 1.A (6,9)-graph H.

Also, $PC(H; q) = 9p + 20p^2 + 19p^3 - 18p^4 - 62p^5 - p^6 + 128p^7 - 106p^8 + 26p^9$. In general, $PC(G; q)$ can be written as a function of $p$.

$$PC(G; q) = A_1p + A_2p^2 + \dots + A_{m-1}p^{m-1} + A_mp^m \tag{4}$$

In the next section we demonstrate how the coefficients $A_i$ of $p^i$ are determined by the subgraph structure.

First, note that $PC(G; q) = \Sigma\{R_{s,t}(G; q): \{s, t\} \subseteq V(G)\}$. And, using Provan's result in [18] that computing $R_{s,t}(G; q)$ is NP-hard for $G$ planar of maximum degree three, one can derive the following.

THEOREM. Determining $PC(G; q)$ is NP-hard for the case where $G$ is planar of maximum degree four.

An important observation first made by Kelmans [15] is that for global connectivity $R(G1; q)$ and $R(G2; q)$ can cross. Boesch [7] presents an example for two (6,8)-graphs where $R(G2; q) - R(G1; q) = q^2(1 - q)^5(1 - 3q)$. Thus for $q < 1/3$ graph G2 is more reliable, but for $q > 1/3$ graph G1 is more reliable. This motivates the definition that an (n, m)-graph H is a uniformly optimally reliable graph with respect to global reliability if $R(H; q) \geq R(G; q)$ for all $q$, $0 < q < 1$, and all (n, m)-graphs G.

For the globally-connected reliability measure $R(G;q)$, Boesch [7] has conjectured that there always exists a uniformly optimally reliable (n, m)-graph. Note that for globally-connected reliability, all trees on n vertices have the same reliability. However, for pair-connected reliability there exist trees T1 and T2 such that neither is uniformly more reliable than the other. Figure 2 shows trees T1 and T2 and the ranges over which each is more reliable than the other.

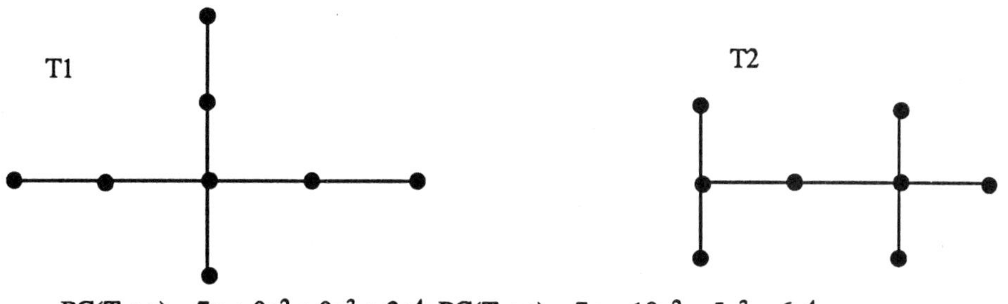


$PC(T_1; q) = 7p + 9p^2 + 9p^3 + 3p^4$; $PC(T_2; q) = 7p + 10p^2 + 5p^3 + 6p^4$
T2 is more reliable than T1 for $0 < p < 1/3$,
T1 is more reliable than T2 for $1/3 < p < 1$.

Figure 2. Trees T1 and T2, neither uniformly more reliable than the other.

For trees the computation of $PC(T; q)$ is straightforward. For an arbitrary graph G the distance distribution of G is $D(G) = (d_1(G), d_2(G), \ldots, d_{n-1}(G))$ where $d_i(G)$ denotes the number of pairs of vertices in G with distance i between them.

THEOREM. The distance distribution D(T) of a tree T completely determines PC(T; q), namely $PC(T; q) = \sum_{i=1}^{n} d_i(T)p^i$.

For any graph G one has $R_{s,t}(G; q) \geq p^{dist(s,t)}$ where dist(s,t) is the distance between s and t, and the next result follows.

THEOREM. For any graph G, $PC(G; q) \geq \sum_{i=1}^{n-1} d_i(G)p^i$.

COROLLARY. For pair-connectivity the star $K_{1,n-1}$ is the uniformly optimally reliable tree on n vertices, and the path $P_n$ is the uniformly least reliable tree on n vertices.

In contrast to the result for trees and the conjectured result for all terminal reliability, the following result is somewhat surprising.

THEOREM. There does not exist a uniformly optimal (n, m)-graph for pair-connected reliability if $n \leq m \leq \binom{n}{2} - 2$.

Efficient algorithms for computing PC(G; q) are described in [1] for trees and series-parallel graphs. And for certain classes of graphs formulas for PC(G; q) have been determined (see [1, 2]). Here formulas are presented just for the n-cycle $C_n$ and the wheel $W_{n+1} = K_1 + C_n$ on n + 1 vertices. Perhaps not surprisingly, even for such simple graphs as the wheels the formulas are quite complicated.

$$PC(C_n; q) = n\,\frac{p-p^n}{q} - \frac{n(n-1)}{2}\,p^n$$

$$PC(W_{n+1}; q) = n\left[1 - \frac{q^3}{(1-pq)^2}\right]\left[1 - (pq)^n\right] - n^2\,\frac{(pq)^{n+1}}{(1-pq)^2}$$

$$+ \frac{n(n-1)}{2}\left[1 - \frac{q^3}{(1-pq)^2}\right]^2$$

$$+\left[\frac{q^4(1+3p^2)+2p^4q^3}{(1-pq)^4}\right]\left[n\,\frac{(pq)-(pq)^n}{1-pq}-\frac{n(n-1)}{2}\,(pq)^n\right]$$

$$+\frac{pq^4}{(1-pq)^3}\left[\frac{-n^2(pq)^n+n^2(pq)^{n+1}+npq-n\,(pq)^{n+1}}{(1-pq)^2}-\frac{n^2(n-1)}{2}\,(pq)^n\right]$$

$$+\left(\frac{pq}{1-pq}\right)^2\frac{n^2(n-1)\,(n+1)}{12}\,(pq)^n$$

3. SUBGRAPH DETERMINATION OF PC(G; q). As in formula (4) we let $PC(G;q) = A_1p + A_2p^2 + \dots + A_{m-1}p^{m-1} + A_mp^m$. In this section we illustrate the next theorem using graph H of Figure 1.

THEOREM. For a fixed value h, the coefficients $A_1, A_2, \dots, A_h$ in PC(G; q) can be computed in time polynomial in n.

First, note that the coefficient $B_i$ in (3) only affects $A_i, A_{i+1}, \dots, A_m$ in (4). Viewed directly, $A_i$ is not affected by any state S with $|S| \geq i+1$. Second, let #(G; H) denote the number of (not necessarily induced) subgraphs of G isomorphic to H. For example, in a tree T for path $P_k$ on k vertices ($P_k$ has length k - 1) one has $\#(T; P_k) = d_{k-1}(T)$. Thus, for a tree T in PC(T; q) we have $A_i = \#(T; P_{i+1})$.

In general, for an (n, m) -graph G and $1 \leq i \leq m$ let $G(i) = \{(H, \{s,t\})$: H is an i-edge subgraph of G with two designated vertices s and t, and each of the i edges is on at least one s-t path$\}$. Then $A_i$ in PC(G; q) equals the sum of the $A_i$ terms in $\{R_{s,t}(H;q)$: $(H, \{s,t\}) \in G(i)\}$.

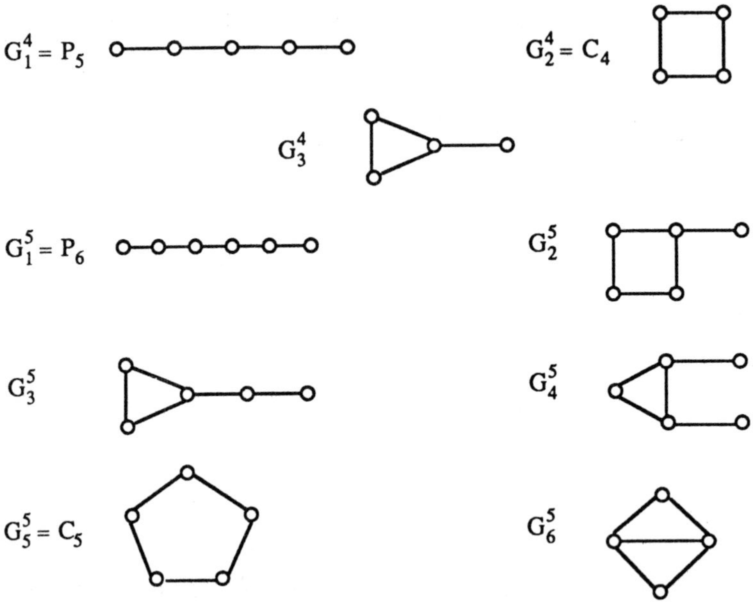


Figure 3. Possible graphs H in G(4) and G(5).

THEOREM. For any graph G, in PC(G: q)

1) $A_1 = \#(G; P_2)$,
2) $A_2 = \#(G; P_3)$,
3) $A_3 = \#(G; P_4) - 3\#(G; C_3)$,
4) $A_4 = \#\left(G; G_1^4\right) - 6\#\left(G; G_2^4\right) - 2\#\left(G; G_3^4\right)$, and
5) $A_5 = \#\left(G; G_1^5\right) - 3\#\left(G; G_2^5\right) - 2\#\left(G; G_3^5\right) - \#\left(G; G_4^5\right)$
$- 10\#\left(G; G_5^5\right) + 7\#\left(G; G_6^5\right)$.

As previously stated, for graph H in Figure 1 we have $PC(H; q) = 9p + 20p^2 + 19p^3 - 18p^4 - 62p^5 - p^6 + 128p^7 - 106p^8 + 26p^9$. Note that $A_1 = 9$, $A_2 = 20$, $A_3 = 31 - 3 \bullet 4 = 19$, $A_4 = 32 - 6 \bullet 3 - 2 \bullet 16 = -18$ and $A_5 = 17 - 3 \bullet 10 - 2 \bullet 18 - 14 - 10 \bullet 2 + 7 \bullet 3 = -62$.

[1] A. T. Amin, K. T. Siegrist and P. J. Slater, On the Expected Number of Pairs of Connected Nodes: Pair-connected Reliability, Computers and Mathematics with Applications, to appear.

[2] A. T. Amin, K. T. Siegrist and P. J. Slater, Exact Formulas for Reliability Measures for Various Classes of Graphs, Proc. 18th Southeastern Conf. on Combinatorics, Graph Theory and Computing (1987) to appear.

[3] A. T. Amin, K. T. Siegrist and P. J. Slater, Pair-connected Reliability of a Tree and Its Distance Degree Sequences, Proc. 18th Southeastern Conf. on Combinatorics, Graph Theory and Computing (1987) to appear.

[4] A. T. Amin, K. T. Siegrist and P. J. Slater, On the Nonexistence of Uniformly Optimal Graphs for Pair-connected Reliability, submitted for publication.

[5] M. O. Ball, Computational Complexity of Network Reliability Analysis: An Overview, IEEE Trans. on Reliability R-35 (1986) 230-239.

[6] F. T. Boesch, Synthesis of Reliable Networks--A Survey, IEEE Trans. on Reliability R-35 (1986) 240-246.

[7] F. T. Boesch, On Unreliability Polynomials and Graph Connectivity in Reliable Network Synthesis, J. of Graph Theory 10 (1986) 339-352.

[8] B. N. Clark, E. M. Neufeld, and C. J. Colbourn, Maximizing the Mean Number of Communicating Vertex Pairs in Series-Parallel Networks, IEEE Trans. on Reliability R-35 (1986) 247-251.

[9] C. J. Colbourn, Network Resilience, Computers and Mathematics with Applications, to appear.

[10] N. Deo, On the Survivability of Communication Systems, IEEE Trans. on Communication Systems CS-12 (1964) 227-228.

[11] H. Frank and I. T. Frisch, Analysis and Design of Survivable Networks, IEEE Trans. of Communication Technology COM-18 (1970) 501-519.

[12] F. Harary and P. A. Ostrand, How Cutting is a Cutpoint?, Combinatorial Structures and Their Application, Gordon and Breach, New York (1970) 147-149.

[13] F. Harary and P. A. Ostrand, The Cutting Center Theorem for Trees, Discrete Math 1 (1971) 7-18.

[14] F. Harary and P. J. Slater, A Linear Algorithm for the Cutting Center of a Tree, Information Processing Letters 23 (1986) 317-319.

[15] A. K. Kelmans, Connectivity of Probabilistic Networks, Automatika i Telemekhanika 3 (1966) 98-116.

[16] T. Politof and A. Satyanarayana, Efficient Algorithms for Reliability Analysis of Planar Networks--A Survey, IEEE Trans. on Reliability R-35 (1986) 252-259.

[17] J. S. Provan and M. Ball, The Complexity of Counting Cuts and Computing the Probability that a Graph is Connected, SIAM J. Comput. 12 (1983) 777-778.

[18] J. S. Provan, The Complexity of Reliability Computations in Planar and Acyclic Graphs, Univ. of North Carolina Tech TR 83/12 (1983).

[19] R. Van Slyke and H. Frank, Network Reliability Analysis I, Networks 2 (1972) 279-290.

[20] R. S. Wilkov, Analysis and Design of Reliable Computer Networks, IEEE Trans. on Communication Com-20 (1972) 660-678.

[21] R. K Wood, Factoring Algorithms for Computing k-Terminal Network Reliability, IEEE Trans. on Reliability R-35 (1986) 269-278.

Department of Mathematical Sciences
Clemson University
Clemson, SC 29634-1907

CURRENT ADDRESS:
Department of Mathematics and Statistics
University of Alabama in Huntsville
Huntsville, AL 35899

Contemporary Mathematics
Volume **89**, 1989

# ON MINIMUM CUTS OF CYCLES AND MAXIMUM DISJOINT CYCLES

Jayme L. Szwarcfiter

ABSTRACT. We describe a new family of digraphs, named connectively reducible, for which we prove that the minimum cardinality of a set vertices intersecting all cycles equals the maximum cardinality of a set of vertex disjoint cycles. In addition, formulate polynomial time algorithms for the problems of recognition and finding these minimum and maximum sets for digraphs of the family. Similar results hold for the currently existing families of fully reducible and cyclically reducible digraphs. Neither the fully reducible are contained nor contain the cyclically reducible. However, we show that the connectively reducible digraphs contain bother of the existing families.

1. INTRODUCTION. Frank and Gyárfás [1] have shown that for fully reducible digraphs the minimum cardinality set of vertices intersecting all cycles equals the maximum cardinality set of vertex disjoint cycles. Furthermore, there are polynomial time algorithms for finding such a minimum set of vertices [4-6] for this family of digraphs, whereas the same problem is well known to be NP-hard in the general case [2-3]. More recently, Wang, Lloyd and Soffa [9] defined another family of digraphs, called cyclically reducible, which also enables the computation of the above sets in polynomial time. In addition, both these families of digraphs can be recognized in polynomial time [8-9]. However, as noted in [9], the fully reducible digraphs neither are contained nor contain the cyclically reducible ones. In the present paper, we define a new family of digraphs, named connectively reducible, and present the following results:

(i) A proof that the above min-max equality is valid for them.

(ii) A polynomial time algorithm which recognizes digraphs of this kind and finds the corresponding minimum and maximum sets, of vertices and cycles, respectively for digraphs of the family.

(iii) A proof that the connectivelly reducible digraphs contain both the fully and cyclically reducible ones.

1980 Mathematics Subject Classification (1985 Revision)
05C20, 05C38, 05C70.

The following is the plan of the paper. In Section 2, we present the concepts of critical vertices and cycles, in which are based the proposed results. These lead to the idea of critical sequences and connectively reducible digraphs, defined in Section 3. A characterization of the proposed family of digraphs is given in Section 4. The min-max theorem is proved in Section 5, whereas in the following we formulate the polynomial time algorithm for finding the minimum and maximum sets. The algorithm is based on the characterization previously described. The proofs that connectivelly reducible digraphs contain cyclically and fully reducible ones are presented in Sections 7 and 8, respectively. Some further remarks form the last section.

Throughout the paper, D denotes a digraph with vertex set V(D) and edge set E(D). If $v \in V(D)$ and $V' \subseteq V(D)$ then D-v and D-V' represent the digraphs obtained from D by removing v and V', respectively. We use the term component meaning a strongly connected component of D. A component is trivial if it consists of a single vertex. T(D) denotes the subset of vertices of D which are trivial components and $\overline{T}(D)=V(D)-T(D)$. A cycle cut or feedback vertex set of D is a subset of vertices, denoted $\alpha(D)$, intersecting all cycles of D. Two cycles which are vertex disjoint are simply called disjoint. The notation $\beta(D)$ represents a set of disjoint cycles. In an acyclic digraph, if there is a path from vertex v to w then v is an ancestor of w, and w a descendant of v; in addition if $v \neq w$ then v is a proper ancestor and w a proper descendant. Finally, we employ the same notation to represent some operations in sets or sequences, the meaning being clear from the context.

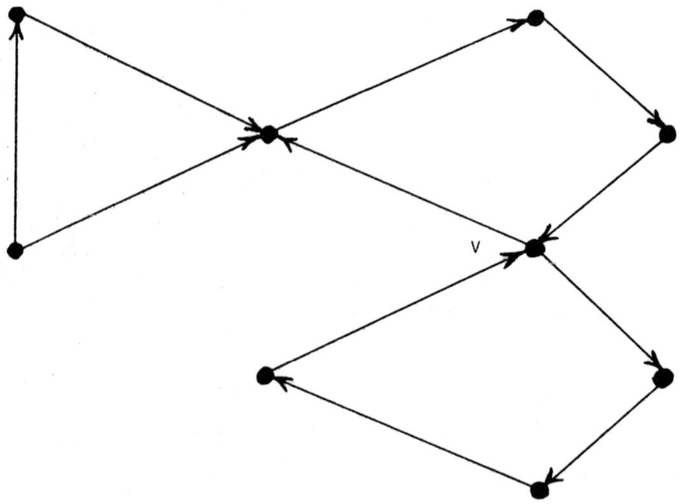


FIGURE 1: A CONNECTIVELY REDUCIBLE DIGRAPH

2. CRITICAL VERTICES AND CYCLES. In this section we present the concept and properties of critical vertices and cycles of a digraph, in which are based the results later described.

Let D be a digraph and v,w vertices of it. The class of v in D is the subset of vertices $\{v\} \cup T(D-v)$, which we denote by $[v,D]$. The classes $[v,D]$ and $[w,D]$ are distinct when $[v,D] \neq [w,D]$.

A vertex $v \in V(D)$ is critical in D when the subgraph induced in D by $[v,D]$ has at least one cycle C. In this case, C is a critical cycle of v in D.

The first lemma relates critical vertices and cycles.

Lemma 1: Let v be a critical vertex and C a critical cycle of v in D. Then C contains v.

Proof: Suppose the contrary. Then there exists a cycle C' formed solely by vertices of some subset of $T(D-v)$. Consequently, every vertex $w \in V(C')$ belongs to a non trivial component of the subgraph induced in D by $\{v\} \cup T(D-v)$. The latter contradicts $w \in T(D-v)$□.

We now describe a condition for two classes to be distinct.

Lemma 2: Let v,w be critical vertices in D. Then $v \in [w,D]$ if and only if $[v,D]=[w,D]$.

Proof: We consider $v \neq w$, otherwise the result is trivial. If $v \in [w,D]$ then v belongs to a trivial component of D-w, that is, every cycle passing through v contains also w. Since v is also critical, there exists a cycle C formed by a subset of trivial components of D-v. By lemma 1, C contains v. That is, w is a trivial component of D-v and then $w \in [v,D]$. Consider now a vertex $z \neq v,w$ such that $z \in [w,D]$. In this case, every cycle C' containing z passes through w. Since $w \in [v,D]$ we conclude that C' also contains v. Then $z \in [v,D]$ and hence $[v,D]=[w,D]$. The converse is immediate, since $v \notin [w,D]$ implies $[v,D] \neq [w,D]$, because $v \in [v,D]$□.

The next lemma asures that any critical cycle contains all critical ver-
ices of its class.

Lemma 3: Let v,w be critical vertices in D such that $[v,D]=[w,D]$. Then a
ycle contains v if and only if it contains also w.

Proof: Suppose there exists in D some cycle C containing v, but not w.
hen C remains a cycle in D-w. Because C contains v, it follows that v can not
e a trivial component of D-w. Consequently, $v \notin [w,D]$. Then we apply lemma 2
nd conclude that $[v,D] \neq [w,D]$, which contradicts the hypothesis. Therefore C
ontains both v and w□.

There are certain vertices which may belong to more than one distinct
lass of a digraph. These vertices satisfy the following condition.

Lemma 4: Let v,w be critical vertices in D such that $[v,D] \neq [w,D]$, and
$\in [v,D] \cap [w,D]$. Then there is no critical cycle in D containing z.

Proof: Suppose the lemma false. Then there is a critical cycle C of v which contains z. Because $[v,D] \neq [w,D]$ we conclude by lemma 2 that $w \notin [v,D]$. Hence $w \notin V(C)$. On the other hand, $v \in V(C)$. Consequently, C remains a cycle in D-w. Since $z \in V(C)$, z can not be a trivial component of D-w, i.e., $z \notin [w,D]$, which contradicts the hypothesis □.

The next lemma describes a condition for two critical cycles to be disjoint.

Lemma 5: Let v,w be critical vertices in D, and C,C' critical cycles of v,w, respectively. Then C,C' are disjoint if and only if $[v,D] \neq [w,D]$.

Proof: Suppose C,C' disjoint and $[v,D] = [w,D]$. In this case, according to lemma 3, both cycles C,C' contain both vertices v,w. Then C,C' are not disjoint, a contradiction. That is, $[v,D] \neq [w,D]$, necessarily. Conversely, when $[v,D] \neq [w,D]$ we apply lemma 4 to conclude that no vertex of C or C' can belong to $[v,D] \cap [w,D]$. Therefore C,C' are disjoint □.

Now, we discuss the effect of removing critical vertices.

Lemma 6: Let v,w be critical vertices in D. Then $[v,D] \neq [w,D]$ if and only if w is critical in D-v.

Proof: If $[v,D] \neq [w,D]$ we must prove that w remains critical after removing v. Let C be a critical cycle of w in D. The idea consists of showing that C is also a critical cycle of w in D-v. Let z be a common vertex of $[v,D]$ and $[w,D]$. By lemma 4, we know that there is no critical cycle of D containing z. That is, $z \notin V(C)$. In addition, since every vertex $z' \in V(C)$ necessarily belongs to $[w,D]$ we conclude that $z' \notin [v,D]$. Therefore, C is preserved in D-v and w remains critical. The converse is simple, as follows. If w is critical in D-v then $w \notin [v,D]$, necessarily. Otherwise, if $w \in [v,D]$ either $w \notin V(D\text{-}v)$ or w becomes a trivial component in D-v. In none of these cases can w be a critical vertex in D-v, a contradiction. Now, when $w \notin [v,D]$ we apply lemma 2 and conclude that $[v,D] \neq [w,D]$ □.

Lemma 7: Let v,w,z be critical vertices in D. Then $[v,D] \neq [w,D]$ if and only if $[v,D\text{-}z] \neq [w,D\text{-}z]$.

Proof: Initially, we consider the hypothesis $[v,D] \neq [w,D]$. If $z \in [v,D]$ then $z \notin [w,D]$. Otherwise, z would be a critical vertex belonging to $[v,D]$ and $[w,D]$, simultaneously; then, by lemma 2, $[v,D] = [z,D]$ and $[w,D] = [z,D]$, i.e. $[v,D] = [w,D]$ a contradiction. Now, $[v,D] = [z,D]$ implies that v can not be a critical vertex in D-z, by lemma 6. Also, $[w,D] \neq [z,D]$ means that v must be a critical vertex in D-z, since no critical cycle of w in D can contain z, according to lemma 5. Therefore, $[v,D\text{-}z] \neq [w,D\text{-}z]$ and the lemma is valid for this case. If $z \in [w,D]$ we apply a similar argument. It remains to analyse the situation $z \notin [v,D],[w,D]$. Suppose the lemma false, that is, $[v,D\text{-}z] = [w,D\text{-}z]$ and let C be a critical cycle of v in D. Then $w \notin V(C)$, since it follows from the hypothesis that $w \notin \lfloor v,D \rceil$. In addition, C must contain some vertex $x \in \lfloor z,D \rceil$, $x \neq z$. Otherwise, C would remain as a critical cycle of v in D-z; and because $\lfloor v,D-z \rceil = \lfloor w,D-z \rceil$ we conclude by lemma 3 that C also contains w, a contradiction. Consequently, in fact $x \in \lfloor z,D \rceil$. In addition, since C is a critical cycle of v in D we know that $x \in \lfloor v,D \rceil$. In the present situation, v and z are two critical vertices in D such that $\lfloor v,D \rceil \neq \lfloor z,D \rceil$ and x is a common vertex of $\lfloor v,D \rceil$ and $\lfloor z,D \rceil$. By lemma 4, we can see that there is no critical cycle in D containing x. Therefore, C does not exist, which contradicts the fact that v is a critical vertex. Consequently, $\lfloor v,D-z \rceil \neq \lfloor w,D-z \rceil$ and the proof of necessity is completed. Conversely, let the hypothesis $\lfloor v,D-z \rceil \neq \lfloor w,D-z \rceil$. There are four cases to consider:

(i) $z \in \lfloor v,D \rceil, \lfloor w,D \rceil$.

Then by lemma 2, $\lfloor v,D \rceil = \lfloor w,D \rceil = \lfloor z,D \rceil$. In this case, $\lfloor v,D-z \rceil = \lfloor v,D \rceil - \{z\}$ and $\lfloor w,D-z \rceil = \lfloor w,D \rceil - \{z\}$. That is, $\lfloor v,D-z \rceil = \lfloor w,D-z \rceil$, contradicting the hypothesis. Therefore, this case does not occur.

(ii) $z \in \lfloor v,D \rceil$ and $z \notin \lfloor w,D \rceil$.

That is, $\lfloor v,D \rceil \neq \lfloor w,D \rceil$ and the lemma holds.

(iii) $z \notin \lfloor v,D \rceil$ and $z \in \lfloor w,D \rceil$.

Similar to (ii).

(iv) $z \notin \lfloor v,D \rceil, \lfloor w,D \rceil$.

Then $\lfloor v,D \rceil \neq \lfloor z,D \rceil$ and $\lfloor w,D \rceil \neq \lfloor z,D \rceil$. Let C and C' be critical cycles of v and w in D, respectively. By lemma 6, we conclude that v and w remain critical in D-z and therefore C and C' are critical also in D-z. We now apply lemma 5 to D-z and find out that C and C' are disjoint. Next, using again lemma 5, but to the digraph D instead, we finally conclude that $\lfloor v,D \rceil \neq \lfloor w,D \rceil$. This completes the proof □.

3. CRITICAL SEQUENCES. In order to describe the class of connectively reducible digraphs we need the following definitions.

Let D be a digraph and $S=\{v_1,\ldots,v_k\}$ a sequence of vertices of it. The value k is the <u>length</u> of S, while the symbol $S_j$ denotes the subsequence $\{v_1,\ldots,v_j\}$, for any j, $1 \leq j \leq k$. We also write $S_0$ to represent the empty sequence $\emptyset$. The notation $D(S_j)$ means the digraph induced in D by the subset of vertices $\overline{T}(D-S_j)$. Then, for example, $D(S_0)$ is the digraph formed by the non trivial components of D. The digraph $D(S_j)$ is called the <u>resulting</u> of $S_j$. If each vertex v is critical in $D(S_{j-1})$ then S is a <u>critical sequence of</u> D, $1 \leq j \leq k$. In this case, additionally, if D(S) does not contain any critical vertices then S is a complete critical sequence, or simply, complete sequence. Next, a vertex $v \in V(D)$ is strongly non critical if there is no critical sequence of D containing v. Finally, D is connectively reducible when the subgraph induced in it by the subset of all strongly non critical vertices is acyclic.

For example, the digraph of figure 1 has only one critical vertex, namely v. In addition, {v} is its only critical sequence, while the removal of this vertex destroys all cycles. Therefore, it is connectively reducible.

Next, we establish relations between critical vertices and resulting digraphs.

Lemma 8: Let D be a digraph, S a critical sequence of it and v,w critical vertices in D such that $[v,D]=[w,D]$. Then $v \in V(D(S))$ implies:

(i) $[v,D(S)] = [w,D(S)]$,

(ii) $w \in V(D(S))$, and

(iii) v,w remain critical vertices in D(S).

Proof: Let $S=\{v_1,\dots,v_k\}$. We use induction in k. If k=0 the lemma is trivially true. When k>0, assume it valid for all critical sequences of length at most k-1. Let $v \in V(D(S_k))$. Then $v \in V(D(S_{k-1}))$ and we can apply the induction hypothesis to conclude that

(i)' $[v,D(S_{k-1})] = [w,D(S_{k-1})]$,

(ii)' $w \in V(D(S_{k-1}))$, and

(iii)' v,w are both critical in $D(S_{k-1})$.

We can now observe that vertices $v,w,v_k$ are all critical in $D(S_k)$. Therefore, we can apply (i)' to lemma 7 and conclude that $[v,D(S_k)]=[w,D(S_k)]$, because $\overline{T}(D(S_k))=\overline{T}(D(S_{k-1})-v_k)$. This proves (i). Next, since $v \in V(D(S_k))$ we can apply (i) to write $[v,D(S_k)]=[w,D(S_k)]$, which leads to $w \in V(D(S_k))$ assuring (ii).

Finally,

$$[v,D(S_{k-1})] \neq [v_k,D(S_{k-1})],$$

otherwise $v \notin V(D(S_k))$, a contradiction. Therefore, we can apply lemma 6 to obtain that v is critical in $D(S_k)$. Similary for w. The proof of (iii) is now completed □.

We now introduce the concept of representatives of D.

Let D be a digraph and $R(D)=\{v_1,\dots,v_k\}$ some subset of critical vertices of it. R(D) is a critical representative subset, or simply a representative, of D when the following conditions are both satisfied:

(i) $i \neq j \Rightarrow [v_i,D] \neq [v_j,D]$.

(ii) $w \in V(D)$ is a critical vertex of $D \Rightarrow [v_i,D]=[w,D]$, for some $i$, $1 \leqslant i \leqslant k$

In other words, a representative of D is a maximum cardinality subset formed by critical vertices belonging to distinct classes of D.

The next lemma shows a relation between representatives and critical sequences of a digraph.

Lemma 9: Let S be a sequence formed by vertices of a representative of D, in any arbitrary order. Then S is a critical sequence of D.

Proof: Let $S=\{v_1,\ldots,v_k\}$. We employ induction in k. If $k=0$ there is nothing to prove. Otherwise, suppose the lemma holds for all sequences of length at most $k-1$. Since the vertices of S belong to a representative of D we know that each $v_i$ is critical in $D(S_0)$ and $[v_i,D] \neq [v_j,D]$, $i \neq j$. Consequently, we can apply lemma 6 to conclude that $v_k$ is critical in $D(S_1)$. In addition, it follows from lemma 7 that $[v_k,D(S_1)] \neq [v_j,D(S_1)]$, for $1 \leqslant j < k$. Repeating iteratively this argument it results that v is still critical in $D(S_{k-1})$. In this situation, we can use the induction hypothesis to conclude that $\{v_1,\ldots,v_k\}$ is a critical sequence of D □.

4. CHARACTERIZATION OF CONNECTIVELY REDUCIBLE DIGRAPHS. Consider solving the recognition problem for connectively reducible digraphs. A first idea might be to apply the definition and recognize as a member of this family every digraph D whose subgraph induced by its strongly non critical vertices is acyclic. To use this strategy we would need previously to devise a method for finding the set of all strongly non critical vertices. It seems difficult to solve the latter problem directly from the definition, since to identify these special vertices we would need to generate all possible complete sequences of D, whose number can grow exponentially with $|V(D)|$. In this section, we prove a convenient characterization for this family, which enables to recognize connectively reducible digraphs after constructing just one complete sequence.

Theorem 1: All complete sequences of a digraph D have the same resulting digraph.

Proof: Let S be an arbitrary complete sequence. The proof consists of defining a canonical sequence S' and showing that $D(S)=D(S')$, as below detailed. We start by S'.

Constructing S': Let R(D) be a representative of D. A canonical sequence S' of D is recursively defined as follows. If $R(D)=\emptyset$ then $S'=\emptyset$. Otherwise, S' is formed by the vertices of R(D), in an arbitrary order, followed by a canonical sequence of D-R(D).

To verify that the above construction always finds a complete sequence of

D, we use induction in the length k' of S'. If k'=0 the result is correct, since $S'=\emptyset$ means $R(D)=\emptyset$ and there can be no critical sequence without critical vertices. Otherwise, assume the construction correct for lengths at most k-1. From the definition, we know that S' is formed by the vertices of R(D), followed by a canonical sequence of D-R(D), which we now denote by S". The leading vertices of S', i.e. R(D), form a critical sequence of D, according to lemma 9. On the other hand, using the induction hypothesis we conclude that S" is a complete sequence of D-R(D). At this point we can apply the definition and asure that S' is a complete sequence of D, which proves the correctness of the above construction.

To show that D(S)=D(S') the idea consists of transforming S into S' through the application of some different operations. Each operation can result in alterations in S. In this case, we must guarantee that the resulting digraph of the sequence remained the same. If we asure the invariance of D(S) through the process we obtain D(S)=D(S'), which would prove the theorem.

We use four different operations to transform S. Two of them replace certain vertices of S by others, while the remaining operations simply change the order of the sequence.

Now, we describe the transformation from S into S', together with the proofs of invariance of D(S) in the process. The current sequence S is denoted by $\{v_1,\ldots,v_k\}$, while R(D) is precisely the representative of D which S' contains.

Operation 1: For each vertex $v \in R(D)$ verify if S contains some critical vertex $w \in [v,D]$. In the affirmative case, replace w by v in S.

The proof that S is mantained complete and D(S) preserved after the end of the above operations is simple. Let v,w according to the hypothesis, that is, $v \in R(D)$, $w \in [v,D]$, both critical vertices in D and $w=v_i$ for some i, $1\leq i\leq k$. Then $w \in V(D(S_{i-1}))$ and applying lemma 8 we conclude that $v \in V(D(S_{i-1}))$ and that v,w are both critical vertices in $D(S_{i-1})$, belonging to a same class in it. Therefore, by lemma 2, $[v,D(S_{i-1})]=[w,D(S_{i-1})]$. Hence, $D(S_{i-1})-v$ and $D(S_{i-1})-w$ coincide. That is, S is mantained complete and D(S) preserved after each of the vertex replacements.

After operation 1, S may not contain yet all vertices of R(D). The transformation to include in S the remaining desired vertices is given below.

Operation 2: For each $v \in R(D)-S$, determine the value $j\geq 1$ such that v is a critical vertex in $D(S_{j-1})$, but not in $D(S_j)$ and next replace $v_j$ by v in S.

We now describe the proof of correctness of operation 2. We need to show that the new sequence S contains R(D), after all transformations. The argument is inductive. If $R(D)-S=\emptyset$ there is nothing to prove. Otherwise, choose $v \in R(D)-S$. Operation 2 identifies the value $j\geq 1$ satisfying:

$v \in V(D(S_{j-1})) - V(D(S_j))$.

We need to asure that such value j does exist. Since S is complete, $D(S_k)$ does not contain critical vertices. Because $v \in R(D)$, it follows that v is critical in $D(S_0)$. Thus, there exists necessarily j, $1 \leq j \leq k$, such that v is critical in $D(S_{j-1})$, but not in $D(S_j)$. There are two alternatives to consider, namely $v \in V(D(S_j))$ or not. In the first case, v is non critical in $D(S_j)$, by hypothesis. However, this can not occur. Because $v_j$ and v are both critical in $D(S_{j-1})$ and $v_j \notin V(D(S_j))$, it follows that $[v,D(S_{j-1})] \neq [v_j,D(S_{j-1})]$. Hence, by applying lemma 6 we would conclude that v remains critical in $D(S_j)$, a contradiction. Therefore, the only possibility is $v \notin V(D(S_j))$. In this case, using again that $v_j$ and v are critical in $D(S_{j-1})$ and lemma 6, we obtain $[v,D(S_{j-1})] = [v_j,D(S_{j-1})]$. That is,

$$D(S_j) = D(S_{j-1}) - [v,D(S_{j-1})] = D(S_{j-1}) - [v_j,D(S_{j-1})].$$

Hence, after replacing $v_j$ by v, S is still complete. Furthermore, for any vertex $w \in R(D)$, necessarily $w \neq v_j$. Because, if $w=v_j$ then $[v,D(S_{j-1})]=[w,D(S_{j-1})]$. In this case, applying succesively lemma 7 would lead us to $[v,D]=[w,D]$, which contradicts $v,w \in R(D)$. Therefore, each replacement of $v_j$ by v in S increases by one the number of vertices of R(D) which appear in S. This completes the proof of operation 2.

After operations 1 and 2, S necessarily contains R(D). However, in order to transform S into S' we need the vertices of R(D) to appear in the leading positions of S. This is accomplished by the following.

Operation 3: If S contains some vertex $v_j \in R(D)$ such that $v_{j-1} \notin R(D)$, $j>1$, then interchange the positions of $v_j$ and $v_{j-1}$ in S. Repeat the operation until no such $v_j \in R(D)$ exists in S.

The proof of correctness of operation 3 consists in showing that after the last interchange of positions, S is still a complete sequence and that D(S) was preserved. In addition, the $|R(D)|$ leading vertices of the transformed sequence are precisely those of R(D). The argument is again inductive. If S is formed solely by vertices of R(D) there is nothing to prove. Otherwise, for each $v_i \in S-R(D)$ define displacement $(v_i)$ as the number of vertices of R(D) which are at the right side of $v_i$ in S. If displacement $(v_i)=0$ for all $v_i \in S-R(D)$ then operation 3 is not performed and its correctness follows trivially. Otherwise, S contains necessarily a vertex $v_j \in R(D)$ such that $v_{j-1} \notin R(D)$, $1<j\leq k$. In this case, $v_j$ is critical in $D(S_{j-1})$, and clearly also in $D(S_0)$. On the other hand,

$$[v_j,D(S_{i-1})] \neq [v_i,D(S_i)], \text{ for all } 1 \leq i < j.$$

Because, otherwise, if for some i vertices $v_i$ and $v_j$ belong to a same class in

$D(S_{i-1})$ then according to lemma 2 $v_j \notin V(D(S_i))$, which contradicts $v_j \in V(D(S_{j-1}))$. Similary, we conclude that $v_j$ is critical in $D(S_{i-1})$, $1 \leq i < j$. Consequently, $v_{j-1}$ and $v_j$ are both critical vertices and belonging to distinct classes in $D(S_{j-2})$. Now, we apply lemma 6 to certify that $v_{j-1}$ is also critical in $D(S_{j-2})-v_j$. Therefore, we can interchange the positions of $v_{j-1}$ and $v_j$ in S and asure that the new sequence so obtained is still critical and complete. Besides, D(S) is also preserved. Because $D(S_j)$ in both sequences, old and new, equals the digraph obtained by removing the trivial components of $D(S_{j-2}) - \{v_{j-1}, v_j\}$. On the other hand, the change of positions between $v_{j-1}$ and $v_j$ asures that displacement $(v_{j-1})$ decreases by one unit. This completes the proof of correctness of operation 3.

The leading vertices of S are now exactly these of R(D). However, we need them in S with the same ordering as they are in S'. This is the purpose of the last operation below.

Operation 4: Reorder the vertices of R(D) in S, so as to obey the same ordering as they appear in S'.

The correction of it is simple. The sequence formed in S by the vertices of R(D) in its new ordering is itself critical, according to lemma 9. Besides, $D(S_{|R(D)|})$ is the digraph obtained by removing the trivial components of D-R(D). Therefore, the new sequence S is also complete and D(S) is mantained.

Consider now the sequence S after all above operations and let us complete the transformation from S into S'. In both sequences the R(D) leading vertices coincide, respectively. Now, remove R(D) both from S and S'. If $D\text{-}R(D)=\emptyset$ then S=S'. Otherwise, S-R(D) is a complete sequence of D-R(D). Also, S'-R(D) is a canonical sequence of it. In addition, D(S)=D(S-R(D)) and D(S')=D(S'-R(D)). Next, apply the four described operations to S-R(D) which would transform it into a new sequence having the same leading $|R(D-R(D))|$ vertices as S'-R(D), while preserving its resulting digraph. Then remove R(D-R(D)) from both S-R(D) and S'-R(D) and so on iteratively. We can then conclude that any arbitrary complete sequence of D has the same resulting digraph as the canonical one. This completes the proof of theorem 1 .

The next propositions follow directly from the above proof.

Corollary 1: All complete sequences of a digraph have the same length.

Corollary 2: Let D be a digraph and S an arbitrary complete sequence of it. Then D is connectively reducible if and only if $D(S)=\emptyset$.

5. THE MIN-MAX THEOREM. Theorem 2: Let D be a connectively reducible digraph. Then min $|\alpha(D)|=\max|\beta(D)|$.

Proof: If D does not contain critical vertices then all its vertices are strongly non critical. In this case D is necessarily acyclic and the theorem is trivial. Otherwise, let $S=\{v_1,\ldots,v_k\}$ be a complete sequence of D, $k \geqslant 1$. Define the subsets of vertices $\alpha(D)=\{v_1,\ldots,v_k\}$ and cycles $\beta(D)=\{C_1,\ldots,C_k\}$, where $C_j$ is a critical cycle of $v_j$ in $D(S_{j-1})$, $1 \leqslant j \leqslant k$. First, we show that $\alpha(D)$ is a cycle cut of D. Since D is connectively reducible, $D(S_k)=\emptyset$, according to corollary 2. Consequently, for any cycle C of $D(S_0)$ there exists an index j, $1 \leqslant j \leqslant k$, such that C is a cycle in $D(S_{j-1})$, but not in $D(S_j)$. Therefore C contains some vertex $w \in [v_j, D(S_{j-1})]$. Suppose that C does not contain $v_j$. Then $w \notin T(D(S_{j-1})-v_j)$, that is $w \notin [v_j, D(S_{j-1})]$, a contradiction. Hence, C contains $v_j$ and we conclude that $\alpha(D)$ is in fact a cycle cut. Next, we examine $\beta(D)$. Suppose there exists a pair of distinct cycles $C_p$, $C_p \in \beta(D)$ containing a common vertex z. Without loss of generality, let $p<q$. Then $z \notin V(D(S_p))$, because $z \in \{v_p\} \cup T(D(S_{p-1})-v_p)$. That is, $z \notin \overline{T}(D-S_i)$, $i \geqslant p-1$, which contradicts $z \in V(D(S_{q-1}))$ and $z \in V(C_q)$. Therefore, $C_p$, $C_q$ can not contain common vertices. Hence, $\alpha(D)$ and $\beta(D)$ are respectively a cycle cut and a set of vertex disjoint cycles of D, having the same cardinality. Therefore the first is minimum and the second maximum □.

6. THE ALGORITHM. A polynomial time algorithm for recognizing connectively reducible digraphs and finding minimum cycle cuts and maximum sets of disjoint cycles for digraphs of this family is a direct consequence of corollary 2 and theorem 2.

The algorithm below accepts as input an arbitrary digraph D and computes one of the following alternative results. Either it confirms that D is connectively reducible and simultaneously exhibits a minimum cycle cut and maximum set of disjoint cycles, or it reports that D is not connectively reducible.

In the <u>initial step</u>, let $i:=0$, define the digraph $D_i:=D$, the sets $\alpha:=\beta:=\emptyset$ and unmark all vertices. In the <u>general step</u>, if there are no unmarked vertices the process terminates (D is connectively reducible iff $D_i$ is acyclic; in the affirmative case, $\alpha$ and $\beta$ are respectively a minimum cycle cut and maximum set of disjoint cycles of D). Otherwise, choose any unmarked vertex v, mark it and construct class $[v, D_i]$. Next, verify if the subgraph induced in $D_i$ by the vertices of $[v, D_i]$ contains some cycle C. If it does contain, then include v in $\alpha$, include C in $\beta$, define $D_{i+1}:=D_i-[v,D_i]$, unmark all vertices of $D_{i+1}$ and increase i by 1. In any case, repeat the general step □.

There is no difficulty to implement this algorithm in $O(n^2(n+m))$ time, $n=|V(D)|$ and $m=|E(D)|$.

7. CONNECTIVELY AND CYCLICALLY REDUCIBLE DIGRAPHS. In this section we show that the family of connectively reducible digraphs contains the cyclically re-

ducible ones. We start by presenting the definitions of the latter.

Let D be a digraph. A vertex $w \in V(D)$ is blocked in D if there exists a path in D from w to some vertex $z \in \overline{T}(D)$. The associated digraph A(v,D) of D relative to v is the subgraph induced in D by the subset of V(D) that contains v and all vertices that are not blocked in D. A W-sequence of D is a sequence of vertices $\{v_1,\ldots,v_k\}$ such that there are cycles in each of the associated digraphs $A(v_i,D_{i-1})$, $1 \leq i \leq k$, where $D_0 = D$ and

$$D_i = D_{i-1} - V(A(v_i,D_{i-1})).$$

In addition, if $D_k$ is acyclic then the W-sequence is complete. Finally, a cyclically reducible digraph is precisely one that admits a complete W-sequence.

The following lemma relates the above associated digraphs and classes as defined in Section 2.

Lemma 10: Let D be a digraph and $w \in V(D)$. If w is a vertex of A(v,D) then w belongs to $[v,D]$.

Proof: If w is a vertex of A(v,D)then w=v or w is not blocked in D-v. In the first case, the lemma holds. Consider then $w \neq v$. By definition, there exists no path in D-v from w to some vertex $z \in \overline{T}(D-v)$. Therefore $w \in T(D-v)$ , otherwise there is a contradiction if we choose z as a vertex located in the same component as w of D-v, and such that $(w,z) \in E(D)$. Therefore, using the definition of class we conclude that $w \in [v,D]$.

Finally,

Theorem 3: Let D be a cyclically reducible digraph. Then D is connectively reducible.

Proof: If D is acyclic the theorem is trivial. Otherwise, D admits a complete W-sequence $S=\{v_1,\ldots,v_k\}$, $k \geq 1$. The proof consists of showing that S is a complete critical sequence of D. The argument is inductive. Suppose the result true for all digraphs admitting W-sequences with fewer than k vertices. Since D is cyclically reducible, $A(v_1,D)$ has some cycle C. By lemma 10, all vertices of C belong to $[v_1,D]$. That is, $v_1$ is critical in D. In addition, the non trivial components of

$$D - V(A(v_1,D))$$

are identical as those of

$$D - [v_1,D],$$

because $v_1$ is critical in D and according to lemma 10

$$V(A(v_1,D)) \subseteq [v_1,D].$$

Therefore, by removing $v_1$ from D and applying the induction hypothesis to $D-v_1$

we conclude that S is a complete critical sequence of D. Furthermore, $D_k$ is acyclic because D is cyclically reducible. Then the resulting digraph D(S) is empty, since

$$\overline{T}(D_i) = \overline{T}(D(S_i)),\ 0 \leq i \leq k,$$

that is, D is connectively reducible □.

8. CONNECTIVELY AND FULLY REDUCIBLE DIGRAPHS. We prove in this section that the connectively reducible contain the fully reducible digraphs.

A flow digraph is a digraph D together with a distinguished vertex $s \in V(D)$, called root, that reaches all the vertices of D. We say that $w \in V(D)$ dominates $v \in V(D)$ when every path in D from s to v contains w. D is fully reducible if every cycle C of D contains some vertex $w \in V(C)$ which dominates all the vertices of C. In this case, we call w a dominator of C and also of D. The edge of C which is directed to the dominator of this cycle is called a back edge.

Theorem 4: Let D be a fully reducible digraph having root s. Then D is connectively reducible.

Proof: Let L be the set of back edges of D. The argument is by induction on $|L|$. If $|L|=0$ then D is acyclic and the theorem is trivial. Otherwise, suppose the result correct for all fully reducible digraphs with fewer than $|L|$ back edges. Let $w \in V(D)$ be a dominator of D located at a maximal distance of s in D-L. That is, in the acyclic digraph D-L no proper descendant of w is a dominator in D. Let C be a cycle containing the back edge (v,w) and $z \in V(C)$, $z \neq w$. Suppose there exists a cycle C' in D such that $z \in V(C')$, but $w \notin V(C')$. Let w' be the dominator of C'. Observe that w does not dominate w' in D, otherwise there would be a path in D-L starting in w and containing w', which contradicts w as a dominator of D at a maximal distance of s in D-L. Hence there exists a path in D from s to w' that does not contain w. Consequently, this path s-w' followed by the path w'-z in C' forms a path originated in the root of D and intersecting C in some vertex other than its dominator w, which contradicts D as fully reducible. Therefore, if $z \in V(C) \cap V(C')$ then necessarily $w \in V(C')$. In this case, every vertex of C becomes a trivial component in D-w. That is, w is a critical vertex in D, and C is a critical cycle of w in D. Removing w from D and taking the non trivial components of D-w we obtain the resulting digraph D({w}). Let S' be a complete critical sequence of D({w}). Note that D({w}) has fewer than $|L|$ back edges, that is, this digraph is connectively reducible according to the induction hypothesis. By corollary 2, we conclude that the resulting digraph of S' in D({w}) is empty. Consequently, the sequence S formed by w followed by S' is a complete sequence in D satisfying $D(S)=\emptyset$. Therefore, D is connectively reducible □.

9. CONCLUSIONS. We have described a new family of digraphs D named connectively reducible and proved that

$$\min|\alpha(D)| = \max|\beta(D)|.$$

The proofs lead to polynomial time algorithms for finding the minimum set of vertices $\alpha(D)$ and maximum of cycles $\beta(D)$. Furthermore, we have also proved that the proposed family of digraphs contains two others for which similar properties hold, namely the fully reducible and connectively reducible digraphs.

Less is currently known about the equivalent problem for edges instead of vertices, regarding reducible digraphs. In fact, it is not known if in a fully reducible digraph the minimum cardinality set of edges intersecting all cycles equals the maximum cardinality set of edge disjoint cycles. Frank and Gyárfás [1] have conjectured that equality also holds in the edge case. Partial results in this direction were reported in [7].

BIBLIOGRAPHY

1. Frank, A., and Gyárfás, A, "Directed graphs and computer programs", In Problèmes Combinatoires et Théorie des Graphes, Colloque Internationaux C.N.R.S., 260, 1976, pp. 157-158.

2. Garey, M.R., and Johnson, D.S., "Computers and intractability: a guide to the theory of NP-completeness", W.H. Freeman, San Francisco, CA, 1979.

3. Karp, R.M., "Reducibility among combinatorial problems", In Complexity of Computer Computations, Miller, R.E., and Thatcher, J.W., eds., Plenum Press, New York, NY, 1972, pp. 85-103.

4. Markenzon, L., "Propriedades e algoritmos para extensões e especializações de grafos de fluxo redutíveis", Ph.D. dissertation, Coord. Prog. Pós-Grad. Eng., Univ. Fed. Rio de Janeiro, Rio de Janeiro, RJ, 1987.

5. Rosen, B.K., "Robust linear algorithms for cutsets", J.Algorithms 3 (1982), 205-217.

6. Shamir, A., "A linear time algorithm for finding minimum cutsets in reducible graphs", SIAM J. Comput 8 (1979), 645-655.

7. Szwarcfiter, J.L., "On a min-max conjecture for reducible digraphs", Tech. Rep. NCE 0186, Núcleo Comp. Elet., Univ. Fed. Rio de Janeiro, Rio de Janeiro, 1986.

8. Tarjan, R., "Testing flow graph reducibility", J. Comput. Syst. Sci 9 (1974), 355-365.

9. Wang, C., Lloyd, E.L., and Soffa, M.L., "Feedback vertex sets and cyclically reducible graphs", J. ACM 32 (1985), 296-313.

NÚCLEO COMPUTAÇÃO ELETR. AND INST. MATEMÁTICA
UNIVERSIDADE FEDERAL DO RIO DE JANEIRO
CAIXA POSTAL 2324
20.001 - RIO DE JANEIRO, RJ - BRASIL

Contemporary Mathematics
Volume **89**, 1989

# GRAPHS AND FINITELY PRESENTED GROUPS

Andrew Vince

**ABSTRACT.** A graph theoretic approach to the Isomorphism Problem in combinatorial group theory is introduced. In particular, algorithms are given that are conjectured to decide, for a large class of presentations, whether the group is trivial. One of these algorithms is related to a strong form of the Poincaré conjecture for 3-manifolds.

1. **INTRODUCTION.** In Section 2 of this paper a graph is introduced as a tool for investigating the Isomorphism Problem in combinatorial group theory. Let X be a set, F(X) the free group on X, R a subset of "words" of F(X) and N(R) the normal closure of R in F(X), i.e. the smallest normal subgroup of F(X) containing R. If a group G is isomorphic to the factor group F(X)/N(R), then G is said to have a presentation by generators X and relations R, and this is denoted $G = \langle X | R \rangle$. If X and R are finite, then G is said to be finitely presented. The *Isomorphism Problem* asks, given two finitely presented groups, does there exist an algorithm to decide whether the groups are isomorphic. Some background is helpful before stating results and conjectures.

Graphs and algorithms have played a crucial role in combinatorial graph theory from its beginnings over a century ago. Concerning graphs we give the two most famous examples.

**Cayley Graph.** In a 1878 paper [4] Cayley constructed the graph that now bears his name. Given a set X of generators for a group G, the

1980 Mathematics Subject Classification. Primary 05C25, 20F05, 20F10.

directed Cayley graph of G with respect to X has vertex set G, and if $v = ux$ then vertices u and v are joined by an edge directed from u to v and labeled x. Figure 1 is the Cayley graph of the symmetric group $S_3$ of permutations of the set {1,2,3} with respect to the generating set $X = \{x,y\}$, where x=(12) and y=(123). Many properties of Cayley graphs

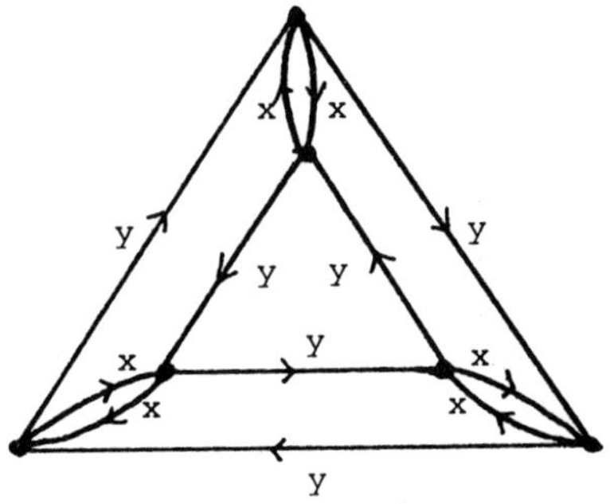


Figure 1. Cayley graph of $S_3$.

have been studied, both by group and by graph theorists. The genus of a group G is defined in terms of this graph; the genus of G is the minimum genus of a surface in which some Cayley graph of G embeds. The theory of ends of groups also depends on the Cayley graph, the number of ends of a group being defined in terms of the number of connected components of the Cayley graph after finitely many edges are removed. Note that in a Cayley graph, edges correspond to generators and each cycle yields a relation in the group.

**Van Kampen Diagram.** Given a group G and a set of generators X, a van Kampen diagram is a directed graph that is (1) planar, (2) has edges labeled by generators in X, and (3) the word assigned to every face, by reading the generators around the cycle, is a relation in G. It follows from this definition that the word corresponding to any cycle in the diagram is a relation in the group. Hence the van Kampen diagram is an aid in deducing new relations from old ones. Consider the group with presentation $\langle x,y \mid xyx=y,\ yxy=x\rangle$. From the outer boundary of the van

Kampen diagram in Figure 2 it can be deduced that $x^4=1$. Loosely, a van Kampen diagram is a portion of a Cayley diagram that can be embedded in the plane.

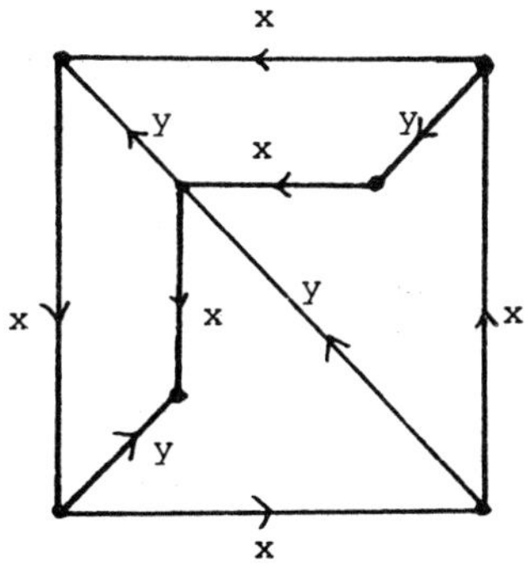


Figure 2. Van Kampen diagram.

Turning our attention from graphs to algorithms, the most famous questions in combinatorial group theory are the *Word Problem* and the *Isomorphism Problem*. In his 1911 paper [5], Dehn asked for a "method of deciding in a finite number of steps" whether a given word in the generators of a finitely presented group is trivial. This is referred to as the Word Problem. In the example of the previous paragraph, a van Kampen diagram sufficed to show that the word $x^4$ is trivial. Note, however, that no algorithmic method was given to produce the van Kampen diagram. In 1908, Tietze [12] posed the Isomorphism Problem: does there exist an algorithm to decide whether two given finitely presented groups are isomorphic? Much of combinatorial group theory and topology developed in parallel (via the fundamental group), so the Isomorphism Problem was likely motivated by questions such as the Homeomorphism Problem: given two topological spaces, are they homeomorphic?

The negative solution to the Word Problem, by Boone and Novikov [2,3,9], did not appear until 1955, after the theory of computability had been developed. Novikov's complicated 143 page proof shows the existence of a finitely presented group with unsolvable Word Problem. Subsequently, many of the combinatorial arguments were unified and the proof was shortened [11]. Using the Boone-Novikov work, Adyan and Rabin [1,10] proved in 1958 that even the following special case of the Isomorphism Problem is unsolvable: Given a finitely presented group, to effectively decide whether the group is trivial. Also in 1958 Markov [8] showed the Homeomorphism Problem unsolvable for piecewise linear manifolds of dimension $\geq 4$. He did this by constructing, for any finitely presented group G, a manifold M(G) whose fundamental group is G. He then showed that G and G' are isomorphic if and only if M(G) and M(G') are homeomorphic, thus reducing the Homeomorphism Problem to the Isomorphism Problem.

In this paper attention is restricted to the following special case of the Isomorphism Problem considered by Adyan and Rabin.

*Triviality Problem*: Given a finitely presented group, does there exist an algorithm to decide whether the group is trivial.

It is the thesis of this paper that, despite the negative result of Adyan and Rabin, such an algorithm exists for a large class of finitely presented groups. Four graph theoretic algorithms are given in Section 3. The required graphs, which are neither Cayley graphs nor van Kampen diagrams, are introduced in Section 2. For each of the four algorithms it is proved that, if the output is "the group is trivial," then indeed the input group is trivial. For Algorithms 1 and 2 there are known examples where the converse fails. That is, finitely presented trivial groups are given which Algorithms 1 and 2 fail to recognize as trivial. However, even Algorithm 1, the simplest of the algorithms, solves the Triviality Problem for certain "2-dimensional" presentations (Theorem 2). Algorithms 3 and 4 solve the Triviality Problem for all examples tried by the author. By the Adyan-Rabin Theorem, however there must exist a counter-example to each of these algorithms. The difficulty is that the group constructed in the Boone-Novikov proof is not given explicitly. Very loosely, the unsolvability of the Word Problem follows from the unsolvability of the Halting Problem for the universal Turing machine and the fact that finitely presented groups can appropriately simulate Turing machines. It would be nice to have explicit counterexamples for Algorithms 3 and 4. The general unanswered question is to determine the class of finitely presented groups for which the algorithms in Section 3 do solve the Triviality Problem. In particular it is conjectured that Algorithms 2, 3, and 4 solve an old open problem - to decide whether a given 3-dimensional manifold is simply connected. More about this conjecture appears in Section 4.

2. **GRAPHS AND PRESENTATIONS.** Graphs are finite and are allowed to have multiple edges but not loops. The vertex and edge sets are denoted V and E, resp. Let $\Gamma$ be a connected graph, C a set of cycles of $\Gamma$ and A a subset of E. The triple $(\Gamma,C,A)$ determines a finite group presentation $\langle X|R\rangle$ as follows. Arbitrarily assign an orientation to each edge in $X = E-A$. For each cycle $c \in C$ let $r_c$ be the word in the symbols X corresponding to a cyclic traversal around c. The exponent of an element $x \in X$ in a word will be +1 or -1 depending on whether the orientation around c is the same or opposite the orientation on x, resp. Let $R = \{r_c | c \in C\}$. The group $G = \langle X|R\rangle$ does not depend on the initial vertex or orientation of a particular cycle; changing the initial vertex or orientation of a cycle c merely gives a conjugate or inverse of the corresponding relation $r_c$. Hence the group $G = \langle X|R\rangle$ is well defined up to isomorphism and will be denoted $G(\Gamma,C,A)$. Now consider the special case where A=T is a spanning tree. Lemma 1 below states that $G(\Gamma,C,T)$ is independent of which spanning tree T is chosen, and therefore the group $G(\Gamma,C,T)$ is simply denoted $G(\Gamma,C)$.

The proofs of Lemma 1 and several subsequent results are simplified by introducing a topological space $K = K(\Gamma,C)$. Let K be the 2-dimensional finite cell complex obtained by attaching a 2-cell to each cycle $c \in C$

along their boundaries. Given a spanning tree T of $\Gamma$, define a map $f: F(X) \to \pi_1(K)$ from the free group on X to the fundamental group of K as follows. Fix a base point b for K. For a given generator $x \in X$, f(x) is the unique cycle in $T \cup \{x\}$, based at b and with direction alligned with that of x. For any word w in the generators, f(w) is defined to make f a homomorphism. Each conjugate of an element of R in F(X) maps to a null homotopic path in K, and hence the same is true for each element of N(R), being a product of such conjugates. In fact, the kernel of f is N(R) and hence $G(\Gamma,C,T) \simeq F(X)/N(R) \simeq \pi_1(K)$. But the fundamental group of $K(\Gamma,C)$ is independent of T, proving the following.

**LEMMA 1.** If $T_1$ and $T_2$ are spanning trees of $\Gamma$, then $G(\Gamma,C,T_1) \simeq G(\Gamma,C,T_2)$. □

Call the pair $(\Gamma,C)$, consisting of a connected graph $\Gamma$ and a set C of cycles of $\Gamma$, a graphical representation of the group G if $G \simeq G(\Gamma,C)$. Let A be a subset of the edge set E of $\Gamma$. An element $x \in E-A$ is said to be dependent on A if there exists a cycle $c \in C$ such that $x \in c \subseteq A \cup \{x\}$. This is analogous to the usual definition of dependence in the cycle matroid of a graph. In Figure 3, for example, edges w and z are dependent on the darkened tree T. For a subset A of E, let $\bar{A}$, the closure of A, be defined by the following algorithm.

**ALGORITHM.** **while** there exists an edge $x \in E-A$ that is dependent on A **do** $A \leftarrow A \cup \{x\}$.

$\bar{A} := A$

**LEMMA 2.** If $A \subseteq B \subseteq \bar{A}$ for subsets A and B of E, then $G(\Gamma,C,A) \simeq G(\Gamma,C,B)$.

PROOF. Let x be an element of E-A that is dependent on A. Then in the presentation of $G(\Gamma,C,A)$, x is a generator. It is a direct consequence of the dependence of x on A that x=1 is a relation. Proceeding inductively, x=1 is a relation for all $x \in \bar{A}-A$ and, in particular, for all $x \in B-A$. Therefore $G(\Gamma,C,B) \simeq G(\Gamma,C,A)$. □

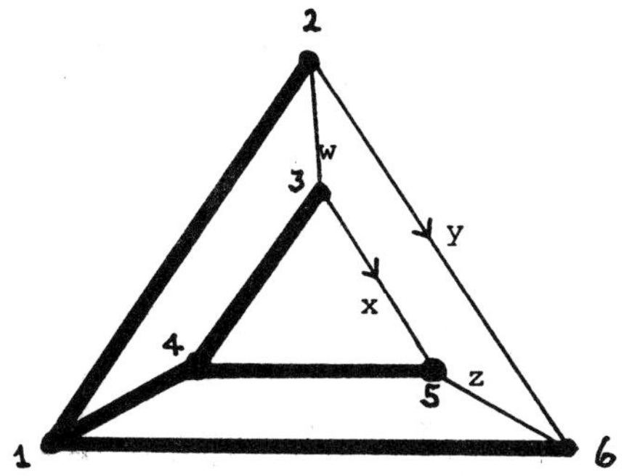


Figure 3. $C = \{(1234),(2356),(1456)\}$.

Using Lemma 2 for the graph in Figure 3, it is easy to check that $G(\Gamma,C)$ is the infinite cyclic group: $\langle x,y \mid x=y\rangle = \langle x \mid \ \rangle$.

**THEOREM 1.** Every finitely presented group has a graphical representation.

PROOF. Let $\langle X \mid R\rangle$ be a finite presentation of the group G. It may be assumed that in each relation there is no repeated generator. For example, if the generator x appears twice in a relation r, then a generator y is added to X and a relation y=x to R and y is substituted for the second occurence of x in r. Now the graph Γ is constructed as follows. Start with a set $E(X) = \{e(x) \mid x \in X\}$ of isolated, directed edges, one for each element of X. For each relation $r = x_1^{\epsilon_1} x_2^{\epsilon_2} \ldots x_n^{\epsilon_n}$, $\epsilon_i=+1$ or $-1$, adjoin a set $E(r) = \{e_1,\ldots,e_n\}$ of edges to E(X) so that $e_1e(x_1)e_2e(x_2)\ldots e_ne(x_n)$ forms the cycle c(r) corresponding to r. If $E(R) = \bigcup\{E(r), r \in R\}$ is connected, then take G(R)=E(R). If E(R) is not connected, then G(R) is obtained from E(R) by adding a vertex v and an edge from v to each connected component of E(R). An example is given in Figure 4 for the free abelian group on two generators $\langle x,y \mid xy=yx \rangle = \langle x,x',y,y' \mid xy=y'x',\ x=x',\ y=y' \rangle$. Let $\Gamma$ consist of all vertices and edges in G(R) and E(X), and let C be the union of $\bigcup\{c(r) \mid r \in R\}$ and all cycles contained in G(R). If T is any spanning tree of $\Gamma$ contained in G(R), then $\overline{T}$ contains G(R). That $(\Gamma,C)$ is a graphical representation of G now follows from Lemma 2. □

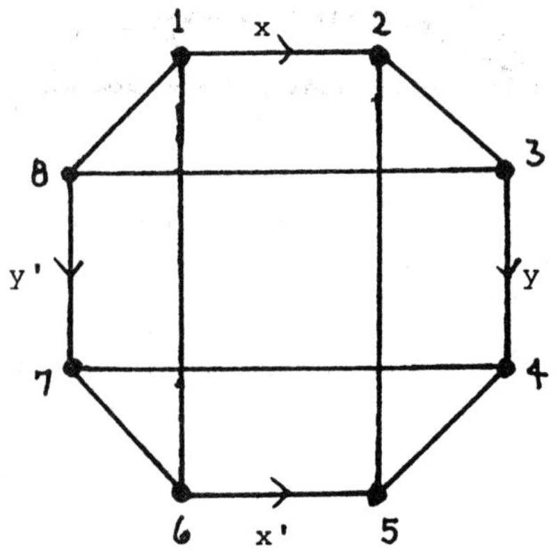


Figure 4. C = {(1256),(3478),(12345678),(16745238)}.

It is sometimes desirable to have a "nice" graphical representation of a group. Corollaries 1 and 2 give two notions of nice. Call a representation $(\Gamma,C)$ <u>non-degenerate</u> if the intersection of any two cycles in C is either empty, a single vertex or a single edge.

**COROLLARY 1.** Every finitely presented group has a non-degenerate graphical representation.

PROOF. Using Theorem 1, let $(\Gamma,C)$ be a graphical representation of the given group G. If cycles c and c' are a degenerate pair, then alter

$\Gamma$ and C as follows. If $c = v_0v_1...v_n=v_0$, then add a vertex v to $\Gamma$ and add cycles $v_1vv_2, v_2vv_3, ..., v_nvv_1$ to C. Remove c from C. Do the same alteration for c'. (In the case of a doubled edge, one edge on each cycle, it may be necessary to first introduce a vertex at the midpoint of each edge of the double.) Repeat the above operation for degenerate pairs of cycles until none remain, resulting in a pair $(\Gamma',C')$. It is clear that the 2-complexes $K(\Gamma,C)$ and $K(\Gamma',C')$ associated with $(\Gamma,C)$ and $(\Gamma',C')$, resp., have the same fundamental group, yielding Corollary 1. □

**COROLLARY 2.** Every finitely presented group has a graphical representation $(\Gamma,C)$ where $\Gamma$ is a trivalent graph.

PROOF. By Theorem 1 the finitely presented group has a graphical representation $(\Gamma,C,T)$. We now show how to alter $\Gamma$ to make it trivalent. Let v be a vertex of $\Gamma$. Three cases are considered.

(1) If deg v=1, then remove v and the edge vu incident to v from $\Gamma$, and remove edge vu from T.

(2) If deg v=2, remove v and the two edges vu and vw incident to v from $\Gamma$. Then add an edge uw. If a cycle in C contains vertices u,v and w consecutively, the cycle is altered to pass through uw. If T contains uv and vw, then replace them by uw in the new spanning tree. If T does not contain both uv and vw, then uw is not included in the new spanning tree.

(3) If deg v>3, then let $vu_1, vu_2, ..., vu_n$, n>3, be the edges incident with v. Remove edges $vu_1$ and $vu_2$. Add a vertex v' and edges $v'u_1$, $v'u_2$ and vv'. Any cycle of C that contains consecutive vertices $u_i, v, u_j$, i=1 or 2, j≥3, is altered to pass through $u_iv'vu_j$. Edge vv' is added to the spanning tree T.

It is easy to check that none of the three alterations changes the group presentation. Steps (1) and (2) remove vertices of degree 1 and

2. Step (3) reduces the degree of a vertex with degree > 3. Therefore repeated use of these steps results in a trivalent graph. □

**REMARK 1.** Corollaries 1 and 2 do not guarantee a graphical representation that is simultaneously non-degenerate and trivalent. If, in the proof of Corollary 2, the construction starts with a non-degenerate graphical representation, then the cycle set C in the construction of Corollary 2 has the property that the intersection of any two cycles is either empty or connected. However, it cannot be concluded that $(\Gamma,C)$ is non-degenerate.

**REMARK 2.** The proofs of Theorem 2 and its corollaries can easily be made into an algorithm for producing the desired graphical representation of a finitely presented group.

3. **THE TRIVIALITY PROBLEM.** In this section the four algorithms mentioned in the introduction are given. In each case the input is a finitely presented group and the first step is to construct a graphical representation as in Section 2. Recall that E denotes the edge set of $\Gamma$. Here $\mathcal{T} = \mathcal{T}(\Gamma)$ denotes the set of all spanning trees of $\Gamma$.

**ALGORITHM 1.** Given a finitely presented group G construct a graphical representation $(\Gamma,C)$ of G

Choose $T \in \mathcal{T}$ and construct $\overline{T}$

**if** $\overline{T}=E$ **then** print "trivial group" **else** print "non-trivial group".

Step 1 uses Remark 2 and Step 2 uses the algorithm of Section 2. It follows from Lemma 2 that if $\overline{T}=E$, then $G(\Gamma,C)$ has presentation $\langle \mid \rangle$ consisting of no generators and relations, and hence is isomorphic to the trivial group. Therefore, if the output of Algorithm 1 is "trivial group," then G is indeed trivial. There are, however, graphical representations of the trivial group that Algorithm 1 does not detect. In the graphical representation of the trivial group in Figure 5a, for example, $\overline{T}=T\neq E$.

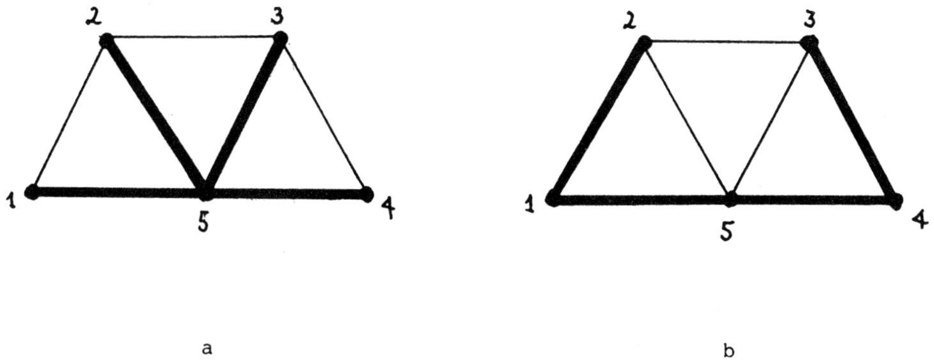


Figure 5. $C = \{(12345),(1235),(2345)\}$.

Algorithm 1 does, however, solve the Triviality Problem for the following special case, which will be discussed further in Section 4.

**THEOREM 2.** If a graphical representation $(\Gamma,C)$ for the given group G is such that each edge in $\Gamma$ is contained in at most two cycles of C, then Algorithm 1 solves the Triviality Problem for G.

PROOF. Assume, by way of contradiction, that $(\Gamma,C,T)$ represents the trivial group, but $\overline{T}\neq E$. Then for any edge $x \in E-\overline{T}$ and any cycle $c \in C$ containing x there exists another edge contained in c and in $E-\overline{T}$. This implies that there is a set F of edges such that every cycle in C contains either 0 or 2 edges of F. Now let $\langle X|R\rangle$ be the presentation determined by $(\Gamma,C,T)$ and let x be any edge in F. Since the group is trivial, x must be the product of conjugates of relations in R or their inverses. But the closed paths in $\Gamma$ corresponding to these relations, as defined prior to Lemma 1 in Section 2, contain an even number of elements of F, and hence so does their product. One the other hand, x contains only 1, an odd number. □

The next algorithm is based on the following observation concerning the example in Figure 5. Although Algorithm 1 fails to recognize the group as trivial using the spanning tree T in Figure 5a, there is another spanning tree T', shown in Figure 5b, for which $\overline{T'}=E$. For this reason we augment Algorithm 1 to try all spanning trees.

**ALGORITHM 2.** Given a finitely presented group G construct a non-degenerate graphical representation $(\Gamma, C)$.

**while** $\mathcal{T} \neq \emptyset$ **do**

Choose $T \in \mathcal{T}$ and construct $\overline{T}$

  **if** $\overline{T}=E$ **then** print "trivial group" and **goto** end

  $\mathcal{T} \leftarrow \mathcal{T}-\{T\}$

**od**

print "non-trivial group"

**end**

Although Algorithm 2 is an improvement over Algorithm 1, there remain simple examples of graphical representations of the trivial group that Algorithm 2 fails to detect. Consider the graph in Figure 6a where the sides of the large boundary triangle are identified in the direction indicated by arrows in Figure 6b. Hence $\Gamma$ has 3 vertices and 8 edges. Here C is the set of six 3-cycles. $G(\Gamma,C)$ can easily be shown trivial, but there is no spanning tree for which $\overline{T}=E$.

Algorithms 3 and 4 that follow are more powerful than Algorithm 2 in the sense that if Algorithm 2 solves the Triviality Problem for a given finitely presented group, then so do Algorithms 3 and 4. We know of no counterexamples to show that either Algorithm 3 or 4 does not solve the Triviality Problem. By the Adyan-Rabin Theorem, such counterexamples must exist, and it would be of interest to find a counterexample to each. Algorithm 3 uses Algorithm 1 repeatedly, augmenting the set C of cycles at each iteration. The following theorem ensures that if Algorithm 3 prints "trivial group" for the input group G, then G is trivial.

**THEOREM 3.** For a graphical representation $(\Gamma,C)$, let T be any spanning tree of $\Gamma$ and $C^*$ the union of C and the set of all cycles in $\overline{T}$. Then $G(\Gamma,C^*) \simeq G(\Gamma,C)$.

PROOF. We have $G(\Gamma,C) \simeq G(\Gamma,C,T) \simeq G(\Gamma,C,\overline{T}) \simeq G(\Gamma,C^*,\overline{T}) \simeq G(\Gamma,C^*,T) \simeq G(\Gamma,C^*)$. The second and fourth isomorphisms are from Lemma 2, and the third is because the closure of T in C is the same as the closure of T in $C^*$. □

**ALGORITHM 3.** Given a finite presentation of a group G construct a graphical representation $(\Gamma,C)$.

$\mathcal{T}_0 \leftarrow \mathcal{T}$

(*) $C_0 \leftarrow C$

**while** $\mathcal{T} \neq \emptyset$ **do**

  Choose $T \in \mathcal{T}$ and construct $\overline{T}$

  **if** $\overline{T}$=E **then** print "trivial group" and **goto** end.

  $C \leftarrow C - \{c \mid c \text{ is a cycle in } \overline{T}\}$

  $\mathcal{T} \leftarrow \mathcal{T} - \{T\}$

**od**

**if** $C \neq C_0$ **then** $\mathcal{T} \leftarrow \mathcal{T}_0$ and **goto** (*)

print "non-trivial group"

**end**

An example where Algorithm 3 detects a trivial representation, but Algorithm 2 does not, is the example of Figure 6 given after Algorithm 2. Starting with the tree in the figure, the cycle consisting of the two edges e and f is added to C after one iteration, and the algorithm ends with $\overline{T}=E$ after two iterations.

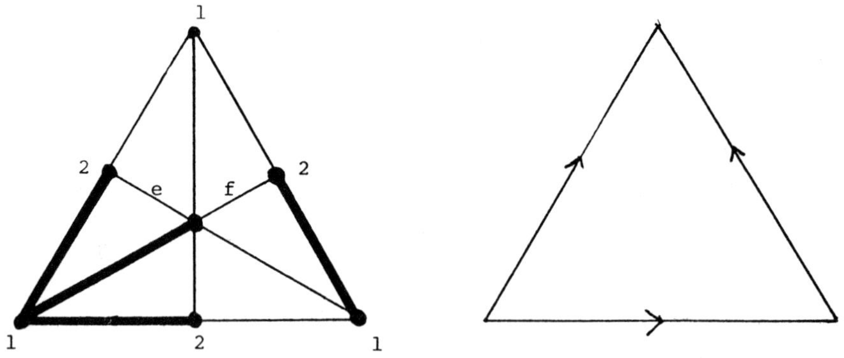


a b

Figure 6.

**REMARK 3.** In certain cases Algorithm 3 actually solves the Triviality Problem. Suppose that, for input group G, Algorithm 3 prints "non-trivial group" at termination. In the terminology of the algorithm let $C' = \{c \in C \mid c \cap (E-\overline{T}) \neq \emptyset\}$ and $\Gamma' = \bigcup\{c \mid c \in C'\}$. For any component $\Lambda$ of $\Gamma'$, if $\Lambda \cap \overline{T}$ is disconnected then G is provably non-trivial. To show this define (1) $\Lambda'=\Lambda$ if $\Lambda \cap \overline{T}$ is connected, and if $\Lambda \cap \overline{T}$ is disconnected then $\Lambda'$ is obtained from $\Lambda$ by adding a vertex $v_o$ and edges from $v_o$ to each connected component of $\Lambda \cap \overline{T}$ and (2) $T'$ is any spanning tree of $\Lambda \cap \overline{T}$ if $\Lambda \cap \overline{T}$ is connected; if $\Lambda \cap \overline{T}$ is disconnected then $T'$ is any spanning tree of the subgraph induced by $\Lambda \cap \overline{T}$ and $v_o$ that includes all the edges incident to $v_o$. Because the presentations are identical, the group $G(\Gamma,C,\overline{T})$ ($\approx G(\Gamma,C,T)$) is isomorphic to the free product of the $G(\Lambda',D,T')$ taken over all the connected components $\Lambda$ of $\Gamma'$, where D is the set of all cycles of C contained in $\Lambda \cap \overline{T}$. Now $G(\Gamma,C)$ is trivial if and only if each factor $G(\Lambda',D)$ is trivial. If $\Lambda \cap \overline{T}$ is disconnected for any component $\Lambda$, then $G(\Lambda',D)$ and hence $G(\Gamma,C)$, is non-trivial. This can be seen by taking a spanning tree $T^*$ of $\Lambda'$ that does not include an edge x incident to $v_o$. Then x is a generator of $G(\Lambda',D,T^*)$ contained in no relation.

The last algorithm uses a product construction. Let $P = (\Gamma, C)$ and $Q = (\Gamma', C')$ be two graphical representations. The product $P \times Q = (\Omega, D)$ is defined as follows. The graph $\Omega$ is the ordinary graph cartesian product of $\Gamma$ and $\Gamma'$ [6]. The set of cycles D is defined by $D = \{c\times\{v'\} \mid c \in C,\ v' \in V(\Gamma')\} \cup \{\{v\}\times c' \mid c' \in C',\ v \in V(\Gamma)\} \cup \{e\times e' \mid e \in E(\Gamma),\ e' \in E(\Gamma')\}$.

**THEOREM 4.** If P and Q are graphical representations of finitely presented groups G and H, then $P \times Q$ is a graphical representation of the direct product $G \times H$.

PROOF. Let T and T' be spanning trees of $\Gamma$ and $\Gamma'$, resp., and $v_o$ an arbitarily vertex of $\Gamma$. Then $T^* = \bigcup\{T\times\{v'\} \mid v' \in V(\Gamma')\} \cup \{v_o\}\times T'$ is a spanning tree of $\Gamma\times\Gamma'$. Since T spans $\Gamma$, the closure $\overline{T^*}$ contains $\bigcup\{\overline{T}\times\{v'\} \mid v' \in V(\Gamma')\} \cup \{\{v\}\times\overline{T'} \mid v \in V(\Gamma)\}$. So if $G=G(P)= \langle X \mid R\rangle$ and $H=G(Q)=\langle Y \mid S\rangle$ with respect to trees T and T', resp., then by Lemma 2, with respect to T*, we have $G(P\times Q) = \langle X \cup Y \mid R \cup S \cup \{xy=yx \mid x \in X, y \in Y\} \rangle = G\times H$. □

In the following algorithm, $K_2$ denotes the graphical representation of the trivial group, where the graph is the complete graph on two vertices and the cycle set is empty. Algorithm 4 applies Algorithm 2 after taking the product with $K_2$.

**ALGORITHM 4.** Given a finite presentation of group G, construct a non-degenerate graphical representation P.

Construct $Q = P\times K_2$; $\mathcal{T} = \mathcal{T}(Q)$

**while** $\mathcal{T} \neq \emptyset$ **do**

  Choose $T \in \mathcal{T}$ and construct $\overline{T}$

  **if** $\overline{T}=E$ **then** print "trivial group" and **goto** end

  $\mathcal{T} \leftarrow \mathcal{T}-\{T\}$

**od**

Print "non-trivial group"

**end**

If Algorithm 4 prints "trivial group", then, by Lemma 2, $G(P \times K_2)$ is trivial and, by Theorem 4, $G=G(P)$ is also trivial. So Algorithm 4 is valid in one direction. Any graphical representation of the trivial group can replace $K_2$, which would result in a potentially more powerful algorithm. In particular, the complete graph $K_n$ can be substituted where all cycles of $K_n$ are included in its cycle set C. However, we have found no examples where anything more complicated than $K_2$ is required. An example where Algorithm 4 detects a trivial representation, but Algorithm 2 does not, is the example $P = (\Gamma, C)$ of Figure 6. A spanning tree of $P \times K_2$ that works is shown in Figure 7.

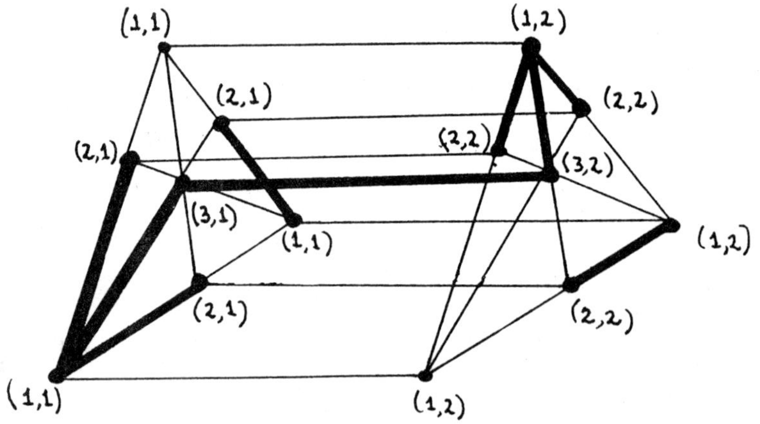


Figure 7. $P \times K_2$.

Although the questions raised in this paper are of more theoretical than practical interest, a brief comment on the complexity of the algorithms is in order. Algorithm 1 is polynomial; the other algorithms are not polynomial because the number of spanning trees in a graph can be exponential in the number of vertices.

4. **RECOGNIZING A SIMPLY CONNECTED 3-MANIFOLD.** A manifold is called closed if it is connected, compact and without boundary. Let M be a closed 3-dimensional manifold given in terms of a finite cell complex.

This is always possible because, in particular, every closed 3-manifold has a finite triangulation. Recall that M is called simply connected if the fundamental group $\pi_1(M)$ is trivial. In this section we consider the following well known open problem.

**PROBLEM.** Given a closed 3-dimensional manifold M, does there exist an algorithm that decides whether M is simply connected?

It is known that the 2-skeleton $M^2$ of M has the property $\pi_1(M) \simeq \pi_1(M^2)$. Therefore we have the following associated graphical representation $(\Gamma,C)$ of $\pi_1(M)$: take $\Gamma$ to be the 1-skeleton of M and C to be the set of boundaries of all 2-cells of M. If any of Algorithms 2, 3, or 4 solve the Triviality Problem for such pairs $(\Gamma,C)$, then we have an affirmative answer to the problem above. This is formulated as the following conjecture.

**CONJECTURE 1.** Let a closed 3-manifold M be given as a finite cell complex. Algorithms 2, 3, or 4, applied to the associated graphical representation, solve the Triviality Problem for the fundamental group of M.

The analogue of Conjecture 1 for closed 2-manifolds (surfaces) is valid. In this case the associated graphical representation $(\Gamma,C)$ of $\pi_1(M)$ has the property that each edge in $\Gamma$ is contained on exactly 2 cycles in C. So $(\Gamma,C)$ satisfies the hypothesis of Theorem 2 and therefore Algorithm 1, applied to $(\Gamma,C)$, decides whether M is simply connected. But this is certainly not new. Surfaces are completely characterized by the Euler Characteristic and orientability, both of which are computable. Hence there is a known algorithm to actually determine the surface up to homeomorphism. The surface, in turn, uniquely determines the fundamental group.

Jumping to 3 dimensions, suppose M is a closed 3-dimensional manifold given in terms of a finite triangulation. If $\Gamma$ is the dual 1-skeleton of M and C is the set of boundaries of 2-cells in the dual 2-skeleton, then $(\Gamma,C)$ is a non-degenerate graphical representation of $\pi_1(M)$ and has the property that each edge of $\Gamma$ is contained on exactly three cycles of C. This condition is analogous to the 2-dimensional case. An affirmative answer to the following question would then imply Conjecture 1.

**QUESTION.** If a non-degenerate graphical representation $(\Gamma,C)$ for a given group G is such that each edge in $\Gamma$ is contained on exactly three cycles of C, then Algorithms 2, 3, or 4 solve the Triviality Problem for G.

Conjecture 2 below is a special case of Conjecture 1. The Poincaré Conjecture states that a simply connected closed 3-dimensional manifold is homeomorphic to the 3-sphere. Conjecture 2 implies the Poincaré Conjecture. This is proved in [14] and indicates that Conjecture 2, if true, is very difficult. It is interesting, nevertheless, to have a completely graph theoretic strong form of the famous conjecture of Poincaré.

**CONJECTURE 2.** Let a closed 3-manifold M be given as a finite cell complex. Algorithm 2, applied to the associated graphical representation, solves the Triviality Problem for the fundamental group of M.

REFERENCES


1. Adian, S. I., The unsolvability of certain algorithmic problems in the theory of groups, Trudy Moskov. Mat. Obsc. 6 (1957), 231-298.

2. Boone, W. W., Certain simple, unsolvable problems in the theory of groups, I, II, III, IV Nederl. Akad. Wetensch. Proc. Ser. A 57 (1954), 231-237, 492-497; 58 (1955), 252-256, 571-577.

3. _____, The word problem, Ann. of Math. (2) 70 (1959), 207-265.

4. Cayley, A., On the theory of groups, Proc. London Math. Soc. 9 (1878), 126-133.

5. Dehn, M., Über unendliche diskontinuierliche Gruppen, Math. Annalen 71 (1911), 116-144.

6. Harary, F., Graph Theory, Reading, Massachusetts, Addison-Wesley, 1972.

7. Lyndon, R. C., and Schupp, P. E., Combinatorial Group Theory. Ergebnisse der Mathematik 84, Berlin-Heidelberg-New York, Springer-Verlag, 1977.

8. Markov, A. A., Insolubility of the problem of homeomorphy, Proc. Internat. Congr. Math. (1958) 300-306.

9. Novikov, P. S., On the algorithmic unsolvability of the word problem in group theory, Trudy Mat. Inst. Steklov 44 (1955), 143.

10. Rabin, M. O., Recursive unsolvability of group theoretic problems, Ann. of Math. (2) 67 (1958), 172-194.

11. Stillwell, J., The word problem and the isomorphism problem for groups, Bull. Am. Math. Soc. (1) 6 (1982), 33-56.

12. Tietze, H., Über die topologischen Invarianten mehrdimensionaler Mannigfaltigkeiten, Montash. f. Math. u. Phys. 19 (1908), 1-118.

13. Vince, A., Generalization of the cycle matroid of a graph, Congressus Numerantium, 40 (1983) 399-407.

14. _____, Graphic matroids, shellability and the Poincaré conjecture, Geometriae Dedicata, 14 (1983), 303-314.



DEPARTMENT OF MATHEMATICS
UNIVERSITY OF FLORIDA
GAINESVILLE, FL 32611

Contemporary Mathematics
Volume **89**, 1989

# PROBLEM CORNER

There was a lot of discussion of open questions by the participants at the conference. Below is a list of those problems presented at the problems session held on Monday afternoon.

## SIX PROBLEMS

Mike Fellows, University of Idaho

### 1. Properties of Sets of Representatives of Vertex Partitions

As far as I know, this problem is due to Frank Hsu in the following form. Suppose $k$ triangles are inscribed in a circle and assume that the points of contact are all distinct. Is it possible to choose one point of contact for each triangle so that no two of the chosen contact points are consecutive with respect to the circle? The conjectured answer is "yes".

Define a partition to be *$k$-thick* (*$k$-thin*) if each class of the partition contains at least (at most) $k$ elements. The *partition independence number* $pi(G)$ of a graph $G$ is the least $k$ such that, for every *$k$-thick* partition of $V(G)$, there is a set of representatives (one vertex from each class of the partition) that is an independent set in $G$.

**Theorem** (Fellows): The following are equivalent:
(1) Hsu's conjecture;
(2) For all $n$, $pi(C_n) = 3$;
(3) For all $n$, $pi(P_n) = 3$;
(4) $pi(P) = 3$, where $P$ is the infinite path.

By Hall's Marriage Theorem, $pi(nC_3) = 3$ and it is easy to work out $pi(K_n)$. How hard is it to compute $pi(G)$? It is not obvious that the problem of determining whether $pi(G)$ is at most $k$ is in $NP$.

The *partition domination number* of a graph $G$ can be defined analogously as the maximum $k$ such that for any *$k$-thin* partition of $V(G)$ there is a set of representatives which is a dominating set in $G$. How difficult is this number to compute for trees?

The original interest of Hsu in this problem arose in the context of studying diameters of Cayley graphs of the integers *mod* $n$. Statement (4) can be interpreted as an attractive conjecture concerning colorings of the integers:

No matter how the integers are colored, if each color is used at least three times, then one can choose one integer of each color so that no two integers differ by one.

### 2. Cutting trees into equal-sized pieces.

A *1-3 tree* is a tree $T = (V, E)$ whose every vertex has degree 1 or 3. Let *cut* $(T)$ denote the minimum number of edges in a subset $E'$ of $E$ such that $T' = T - E'$ consists of two forests, each of order half the order of $T$.

*Proposition:* If $T$ is a 1-3 tree, then *cut* $(T) = O(\log |T|)$.

The problem proposed at the conference of finding a nontrivial lower bound on *cut* $(n) = \max_{|T|=n}$ *cut* $(T)$ has been solved by Fan Chung and Arnie Rosenberg [1], who, coincidentally, presented the result at the conference.

*Problem:* What is the complexity of determining, for input a 1-3 tree $T$ and positive integer $k$, whether *cut* $(T) \leq k$?

1. F.R.K. Chung and A.L. Rosenberg, *Minced trees, with applications to fault-tolerant VLSI processor arrays*, Math. Systems Theory **19** (1986) 1-12.

### 3. Recognizing Cayley graphs

The following problem is easier than graph isomorphism.

*Vertex Transitive*

Instance: A graph $G$.
Question: Is $G$ vertex transitive, i.e. is it the case that for any pair of vertices $u, v$ of $G$, there is an automorphism of $G$ taking $u$ to $v$?

Is the following problem also easier?

*Cayley*

Instance: A graph $G$.
Question: Is $G$ a Cayley graph, i.e., is there a group $A$ and a subset $S \subseteq A$ with $S = S \cup S^{-1}$ so that $G$ is isomorphic to the graph with vertex set $A$ and with an edge $ab$ if and only if $as = b$ for some $s \in S$?

**Theorem** ([1]): A graph $G$ is a Cayley graph if and only if the automorphism group of $G$ has a regular subgroup (that is, the action of the subgroup on $G$ is transitive and without fixed points.)

Is either of *Vertex Transitive* and *Cayley* easier than the other?

1. N. Biggs, Algebraic Graph Theory, Cambridge University Press, 1974.

### 4. Embedding of covers of graphs

A graph $G$ is a *cover* of a graph $H$ if $G$ is a covering space of $H$ for the graphs considered as 1-complexes. An equivalent combinatorial definition is given by the following.

*Proposition:* $G$ is a ($k$-*fold*) cover of $H$ if and only if $G$ is obtained from $H$ by replacing each vertex of $H$ with a set of $k$ vertices and replacing each edge $uv$ of $H$ with a set of $k$ edges $u_i f(u_i), i = 1, \ldots, k$, where $f$ is a bijection between the $k$ vertices $u_1, \ldots, u_k$ replacing $u$ and the $k$ vertices $v_1, \ldots, v_k$ replacing $v$.

One can easily find 2-fold planar covers of $K_{3,3}$ and $K_5$. A covering projection may be thought of as a "local isomorphism". Such maps between graphs have applications to parallel processing networks and to data structures. By the Robertson-Seymour theorems, it is decidable in polynomial time, for any fixed surface $S$, whether a graph $G$ has a cover which embeds on $S$. No constructive proof that this is effectively decidable (in any amount of time or space) is presently known. The difficulty is that there is no known bound on the minimum order of a cover which embeds on $S$ in the case when there is one.

The following conjecture is due independently to Negami and to Archdeacon and Fellows.

*Conjecture:* $G$ has a cover which embeds on $S$ if and only if $G$ has a 2-fold cover which embeds on $S$.

**Theorem** (Archdeacon and Fellows): With possibly the two minimal exceptions $K_{4,4}^-$ and $K_{1,2,2,2}$ in a certain quasiorder that extends the minor ordering, the conjecture is true for the plane.

The techniques employed so far do not generalize to surfaces with higher genus. Archdeacon and Richter have shown that no nonplanar graph has an odd-fold planar cover.

## 5. Binary matroids and matroid minors

Are binary matroids, partially ordered by minors, a Robertson-Seymour poset? That is, can it be shown that:

1. (Well-partial ordering) Any set of matroids has a finite number of minimal elements in the minor ordering.
2. (Polynomial time order test) For any fixed matroid $L$ there is a polynomial time algorithm to determine for an arbitrary matroid $M$ whether $M \geq L$ in the minor ordering.

Paul Seymour has suggested that there may be combinatorial reductions to results on graph minors. It is known that the set of *all* matroids is not well-partially ordered by minors (an infinite antichain can be obtained from the projective planes).

## 6. The complexity of knot triviality and the Arf invariant

The following problem is not in any obvious way decidable.

*Knotlessness*

Instance: A graph $G$.

Question: Can $G$ be embedded in 3-space so that no cycle of $G$ is topologically knotted?

Very little is known about this, except that it is nontrivial.

**Theorem** (Conway and Gordon, Sachs): Every embedding of $K_7$ in 3-space has a knotted cycle.

And solvable in polynomial time.

*Observation:* The set of knotlessly embeddable graphs is closed under minors.

Is the following classical problem as easy?

*Knot Triviality*

Instance: A classical knot projection.

Question: Is the knot topologically trivial?

While *Knot Triviality* is known to be decidable only by a deep and highly exponential time algorithm due to Haken and Schubert, it is presently not known to be $NP$-hard.

In view of the polynomiality of *Knotlessness*, perhaps *Knot Triviality* is also in $P$. Is it in $NP$? (The size of an instance should be measured as the number of crossings in the projection.)

**Theorem** (Reidemeister): Every topologically trivial knot can be combinatorially reduced to the trivial knot (the knot with no crossings) by a sequence of the basic moves:

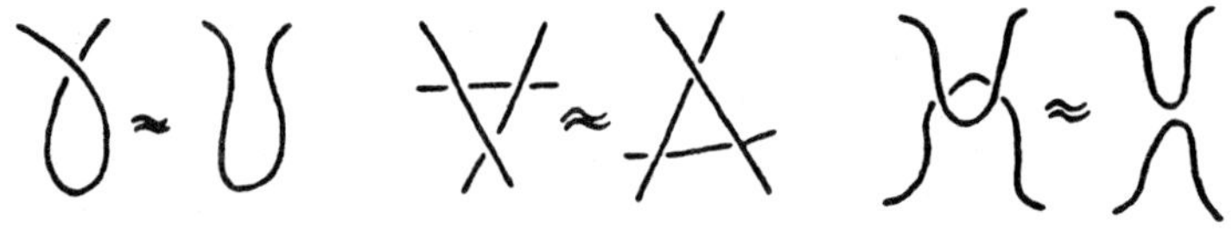

The following problem can be solved in polynomial time by matrix multiplication. The combinatorial interpretation is due to Kauffman.

*Arf Invariant*

Instance: A classical knot projection $K$.

Question: Can $K$ be combinatorially reduced to the trivial knot by a sequence of moves consisting of the Reide-Meister moves and the moves below?

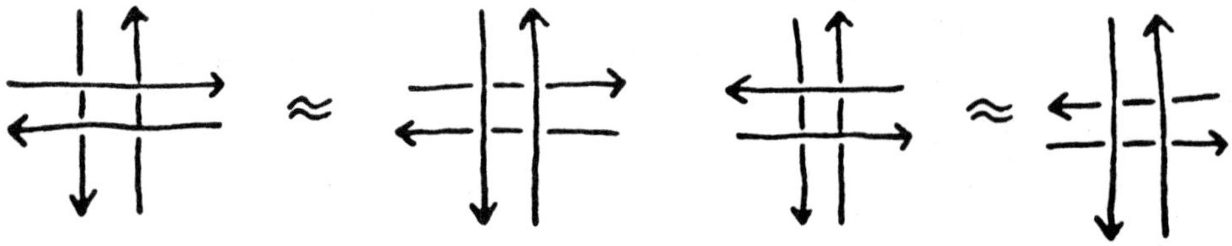

1. K. Reidemeister, Knot Theory, BCS Assocs., Moscow, 1983 .
2. L. Kauffman, Formal Knot Theory, Princeton University Press, 1983.
3. J. Conway and C. Gordon, *Knots and links in spatial graphs,* Journal of Graph Theory **7** (1983) 445-453.
4. H. Schubert, *Bestimmung der primfaktorzerlegung von verkettungen*, Math. Z. **76** (1961) 116-148.

## THE MARTIAN ARTIFACTS PROBLEM

Steven Grantham, Boise State University

Suppose the vertices of the (labelled) complete graph $K_n$ are secretly colored in such a way that a strict majority of the vertices are of a single color. Our task is to identify at least one vertex which we can be sure is of the majority color, but instead of being allowed to test a vertex to see what color it is, we are allowed to ask of any edge of the graph whether its endpoints are of the same or different colors.

Let $f(n)$ denote the minimum number $k$ for which there is an algorithm guaranteed to accomplish this task using no more than $k$ such edge tests. The problem is to compute $f(n)$ and find the algorithm which attains it.

This problem was presented, in much more colorful terms, by Steve Wilson of Northern Arizona University at a conference at Sundance, Utah in August 1985; I am unsure of the original source. In his formulation, we are given a collection of $n$ "Martian artifacts" – delicate globes which are white on the outside but, when broken open, reveal exquisite colors. If two such globes are touched together, they emit an audible signal if and only if they are of the same color. As above, a majority of the artifacts are of one internal color, and we wish to find one of the majority color with as few comparisons as possible.

It seems that most people who have thought about this problem come up with basically the same, fairly natural algorithm, and it is not hard to see that the number of tests, $g(n)$, required by this algorithm in the worst case is given by $g(n) = n-1-B(n-1)$ where $B(n)$ is the number of "ones" in the binary expansion of $n$. It is probably more natural to think of this function as:

$$g(2k-1) = 2k-1-B(2k-1) \text{ and } g(2k) = g(2k-1).$$

The difficult thing, of course, is to verify that this natural algorithm is optimal, i.e. that $f(n) = g(n)$. If the natural algorithm is optimal, then the problem is just as hard with only two colors as it is with an arbitrary number. As far as I know, neither this weaker fact nor the assertion $f(2k) = f(2k-1)$ has been proved.

## UNIQUELY HAMILTONIAN-CONNECTED GRAPHS

Alice M. Dean, Skidmore College

What is the computational complexity of the following?

$UHCV$

Instance: A graph $G$ and a distinguished vertex $v$.

Question: Is $G$ uniquely hamiltonian-connected from $v$?

A graph $G$ is *uniquely hamiltonian-connected from a vertex* $v$ (briefly, $UHC$ from $v$) if, for each vertex $x \neq v$, there is a unique hamiltonian path from $v$ to $x$. Such graphs were introduced by Hendry in [1] and, unlike other variants of hamiltonian graphs, $UHC$ from $v$ graphs have been shown to be very restricted. For example, if the order of $G$ is greater than 3, then all of the following properties hold (cf. [1], [2], [4]):

(1) $n = |V(G)|$ is odd;

(2) $|E(G)| = (3n-3)/2$;

(3) The vertex $v$ is unique, i.e. $G$ is not $UHC$ from any other vertex. Further, $deg(v)$ is even and if $deg(v) = 2t$, then $G$ has exactly $t$ hamiltonian cycles;

(4) The vertices of $G - \{v\}$ all have degree 2,3 or 4; each neighbor of $v$ has degree 3;

(5) $G - \{v\}$ has a unique hamiltonian cycle.

One might hope that these restrictions might make the time complexity of the above decision problem polynomial. On the other hand, the problem $HAM$ : Does $G$ have a hamiltonian cycle? continues to be $NP$-complete even in $G$ is planar, or bipartite with all degrees 2 or 3, etc. [3]. It is not clear that $UHC$ is in $NP$.

$UHCV$ is an element of $D^p$, the set of decision problems which are the intersection of a problem in $NP$ with one in $co-NP$; $UHCV = HCV \cap$ (not $2HCV$). (Definitions of $HCV$, $2HCV$ and related decision problems are given below.)

Observe that $D^p$ contains $NP \cup co-NP$. The following Theorem summarizes what is known about the complexity of these problems. The proofs are either in [3] or can be proved with similar techniques.

**Theorem:**

(1) $HAM$ and $2HAM$ are $NP$-complete; $UHAM$ is in $D^p$ and if it is in $NP$, then $NP = co - NP$.

(2) $HCVX$ and $2HCVX$ are $NP$-complete; $UHCVX$ is in $D^p$ and if it is in $NP$, then $NP = co - NP$.

(3) $HCV$ and $2HCV$ are $NP$-complete.

The goupings of results here suggests the following specific questions.

*Question 1:* Is it true that if $UHCV$ is in $NP$, then $NP = co - NP$?

If the answer to Question 1 is yes, we can ask the following stronger question.

*Question 2:* Is $UHCV$ $D^p$-complete?

*Decision Problems*

$2HAM$ : Instance: A graph $G$.

Question: Does $G$ have two or more hamiltonian cycles?

$UHAM$ : Instance: A graph $G$.

Question: Does $G$ have a unique hamiltonian cycle?

$HCVX$ : Instance: A graph $G$ and vertices $v$ and $x$.

Question: Does $G$ have a hamiltonian path from $v$ to $x$?

$2HCVX$ : Instance: A graph $G$ and vertices $v$ and $x$.

Question: Does $G$ have two or more hamiltonian $v - x$ paths?

$UHCVX$ : Instance: A graph $G$ and vertices $v$ and $x$.

Question: Does $G$ have a unique hamiltonian $v - x$ path?

$HCV$ : Instance: A graph $G$ and a distinguished vertex $v$.

Question: Is $G$ hamiltonian-connected from $v$?

$2HCV$ : Instance: A graph $G$ and a distinguished vertex $v$.

Question: Is $G$ hamiltonian-connected from $v$ *and* are there any vertices $x$ from which $G$ has more than one hamiltonian $v - x$ path?


1. G.R.T. Hendry, *Graphs uniquely hamiltonian-connected from a vertex*, Discrete Math. **49** (1984) 61-74.
2. G.R.T. Hendry, *The size of graphs uniquely hamiltonian-connected from a vertex*, Discrete Math. **61** (1986) 57-60.
3. D.S. Johnson and C.H. Papadimitriou, *Computational complexity*, in The Travelling Salesman Problem, E.L. Lawler and J.K. Lenstra, eds, Wiley, 1985, 37-85.
4. C.J. Knickerbocker, M. Sheard and P.F. Lock, *On the structure of graphs uniquely hamiltonian-connected from a vertex*, in preparation.


## BINOMIAL COEFFICIENT GRAPHS

J.W. Di Paola, New York Academy of Sciences

Consider as vertices the set of all $k$-tuples chosen from a set of $v$ elements. Two $k$-tuples are *adjacent* if they have more than one element in common. This graph is the graph on the binomial coefficient $\binom{v}{k}$ with adjacency coefficient $\lambda = 1$. We symbolize this by $G\binom{v}{k}_{\lambda=1}$.

The problem is to compute the *internal stability number* (or *independence number*) $\alpha$ of this graph. $\alpha$ is the cardinality of a largest set of mutually non-adjacent vertices. It is known that

$$\alpha\left(G\binom{v}{k}_{\lambda=1}\right) \leq \frac{v(v-1)}{k(k-1)}$$

with equality if there exists a balanced incomplete block design with the given $v$, $k$ and $\lambda$ as parameters. Of particular interest are those graphs for which the BIBD does not exist.

What is $\alpha\left(G\binom{36}{6}_{\lambda=1}\right)$?

It is known in this case that $32 \leq \alpha < 42$.

1. J.W. Di Paola, *Block designs and graph theory*, Journal of Combin. Theory 1 (1967).

## PROBLEMS IN ALGORITHMIC GRAPH THEORY

Mark K. Goldberg, R.P.I.

*Definition 1:* Given and integer $n > 0$ and a set $\Gamma$ of graphs on $n$ vertices, a sequence $\{d_1, \ldots, d_n\}$ is *strongly* $\Gamma$ if every graph $\Gamma$ whose vertex degrees are $\{d_1, \ldots, d_n\}$ is in $\Gamma$.

*Definition 2:* A graph is *tough* if, for every subset $A$ of its vertices, the number of connected components of $G - A$ does not exceed $|A|$.

*Definition 3:* A sequence $\{d_1, \ldots, d_n\}$ is *strongly hamiltonian* (resp. tough, connected, 2-factorable) if it is strongly $\Gamma$ where $\Gamma$ is the set of hamiltonian (resp. tough, connected, 2-factorable) graphs.

*Definition 4:* Given a multigraph $G$, the maximum degree and the chromatic index are denoted, respectively, by $\Delta(G)$ and $\chi'(G)$.

*Problem 1:* Is there a polynomial algorithm to determine whether a sequence $\{d_1, \ldots, d_n\}$ is strongly connected (tough, 2-factorable)?

*Problem 2:* Prove there is a polynomial edge coloring algorithm with the following property: given a multigraph $G$, either the coloring constructed by the algorithm is minimal, or the number of colors used does not exceed $\Delta(G) + 1$.

*Problem 3:* Construct a parallel algorithm which runs on a linear number of processors in polylogarithmic time and vertex colors a given graph $G$ in at most $\Delta(G)$ colors when $G$ is not complete or an odd cycle.

*Problem 4:* Construct a parallel algorithm which runs on a linear number of processors in polylogarithmic time and for a given graph $G$ with $n$ vertices and $e$ edges produces an independent set of size $\geq n^2/(2e + n)$.

*Problem 5:* A graph $G(V, E)$ is a *Dirac graph* if the degree of every vertex is at least $|V|/2$. Construct a parallel algorithm which runs on a linear number of processors in polylogarithmic time and, for every Dirac graph, constructs a hamiltonian cycle.

## PRIMAL GRAPH QUESTIONS

Phyllis Chinn, Humboldt State; Bruce Richter, USNA;
and Mirek Truszczynski, Kentucky

Let $G$ be a graph and let $\Gamma$ be a family of graphs. A set $\Phi$ of subgraphs of $G$ is a *decomposition* of $G$ into $\Gamma$ if:

(1) No two graphs in $\Phi$ are isomorphic;
(2) every edge of $G$ lies in exactly one element of $\Phi$; and
(3) each graph in $\Phi$ is isomorphic to a member of $\Gamma$.

The decomposition is *trivial* if $|\Phi| = 1$.

Let $\Omega$ be any class of graphs and let $\Gamma \subseteq \Omega$. Then $\Gamma$ is *primal relative to* $\Omega$ if:

(1) each $G \in \Omega$ has a decomposition into $\Gamma$; and
(2) graphs in $\Gamma$ have only a trivial such decomposition.

These notions were introduced by Dewdney [2] as an analogy to bases of a vector space over $GF(2)$. He proved that the class $\Omega$ of all graphs has a unique subclass $\Pi$ that is primal relative to $\Omega$. A *primal graph* is a member of $\Pi$ and the problem is to characterize the primal graphs.

*Open questions:*

(1) What is the computational complexity of determining whether an arbitrary graph is primal?

(2) So far, all known primal graphs have $\leq 2n$ edges, where $n$ is the order of the graph. Is there a linear bound on the number of edges in a primal graph?

(3) What can be said about graphs which have a unique primal decomposition?

(4) If $\Pi_n$ and $\Omega_n$ are, respectively, the sets of primal and all graphs on $\leq n$ vertices, what is

$$\lim_{n\to\infty} \frac{|\Pi_n|}{|\Omega_n|}?$$

Is it 0?

It is known that $\Pi_n$ contains exponentially (in $n$) many graphs [1].

1. J. Buhler, P. Chinn, B. Richter and M. Truszczynski, *Some results on the number of and complexity of recognizing primal graphs*, to appear.
2. A.K. Dewdney, *Primal graphs*, Aequationes Mathematicae, 4 (1970) 326-328.

## THE TURN CONJECTURE

Dan Sleator, Carnegie Mellon

A *rotation* in a binary rooted tree is a local restructuring of the tree that maintains the symmetric order of the nodes in the tree. A *turn* is a pair of rotations that moves a node two up in the tree. See Figure 1 for a right rotation.

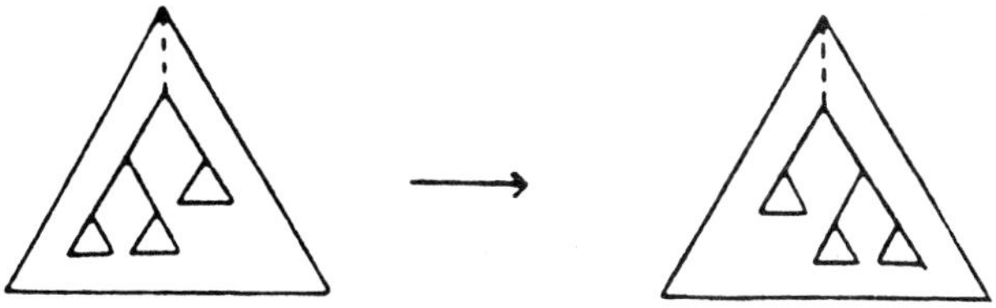

*Turn Conjecture:* Starting from any initial tree of $n$ nodes, at most $O(n)$ right turns can be done.

*Note 1:* It is easy to construct a tree and a sequence of $O(n^2)$ right rotations that can be applied to it.

*Note 2:* There exists an initial $n$-node tree and a sequence of $O(n)$ right rotations and $O(n \log n)$ right turns that can be applied to it.

This problem arises from an attempt to analyse self-adjusting search trees. See [1,2].

1. D. Sleator and R. Tarjan, *Self-adjusting search trees*, JACM **32** (1985) 652-686.

2. R. Tarjan, *Sequential splaying takes linear time*, to appear in Combinatorica.

## OBSTRUCTIONS FOR THICKNESS

Eric Regener, Concordia University

The *thickness* of a graph $G$ is the least $k$ for which $G$ can be partitioned into $k$ planar graphs.

*Question:* What are the minimal graphs having thickness 3? In other words, what are the minimal forbidden subgraphs for a graph to have thickness 2? $K_9 - e$ is one such.

What about the minimal forbidden subgraphs for thickness $k > 2$?

## PACKING TREES INTO COMPLETE BIPARTITE GRAPHS

Arthur Hobbs, Texas A& M

Graphs $G_1$, $G_2, \ldots,$ $G_n$ are *packed into* graph $H$ if the edges of $H$ are colored with colors $0, 1, \ldots, n$ so that each edge of $H$ has exactly one color and for each $i > 0$, the subgraph of $H$ induced by the edges colored $i$, together with additional vertices of $H$ if necessary, is isomorphic to $G_i$.

*Conjecture:* Let $T_2, \ldots, T_n$ be trees such that $T_i$ has order $i$ for each $i$. Then these trees can be packed into $K_{n-1,\lceil n/2 \rceil}$ and into $K_{n,\lceil (n-1)/2 \rceil}$.

Note that $T_2, \ldots, T_n$ have a total of $n(n-1)/2$ edges, so that $K_{n-1,\lceil n/2 \rceil}$ has exactly the necessary number of edges if $n$ is even, and $K_{n,\lceil (n-1)/2 \rceil}$ has exactly the needed number for $n$ odd.

This conjecture has been verified in the following cases [1,3]:

(1) by exhaustion for $n \leq 6$;
(2) in the case each $T_i$ is either a path or a star.

Finally, the following result is known.

**Theorem:** If $T_i$ and $T_j$ are trees having orders $i < j \leq n$, then $T_i$ and $T_j$ can be packed into $K_{n-1,\lceil n/2 \rceil}$.

1. A.M. Hobbs, *Packing trees*, Congress. Numer. **33** (1981) 63-73.
2. A.M. Hobbs, B.A. Bourgeois and J. Kasiraj, *Packing trees in complete graphs*, to appear in Discrete Math.
3. S. Zaks and C.L. Liu, *Decomposition of graphs into trees*, Congress. Numer. **19** (1977) 643-654.

## EDGE-GRACEFUL GRAPHS

Sin-Min Lee, San Jose State University

Let $G = (V, E)$ be a graph and let $f : E \to \mathbb{Z}$ be any function. The *induced map* $f^+ : V \to \mathbb{Z}$ is defined by $f^+(v) = \sum_{uv \in E} f(uv) \ (mod|V|)$.

The graph $G$ is *edge-graceful* if there is a bijection $f : E \to \{1, 2, \ldots, |E|\}$ such that the induced map $f^+$ is a bijection from $V$ to $\{0, 1, \ldots, |V| - 1\}$.

*Conjecture 1:* Any tree of odd order is edge-graceful.

*Conjecture 2:* Any unicyclic graph of odd order is edge-graceful.

*Problem 1:* Which $K_{m,n}$ are edge-graceful?

Lo Sheng-Ping showed that if $G$ is edge-graceful of order $p$ with $q$ edges, then $p$ divides $q^2 + q - p(p-1)/2$.

*Problem 2:* Is this condition sufficient for connected graphs?

Lo Sheng-Ping's condition is not sufficient for disconnected graphs. The simplest example is $C_3 \cup C_4$.

## ON USING $C$-MINORS

Larry Basenspiler, Northern Illinois

Let $G$ and $H$ be simple graphs. $H$ is a *C-minor* of $G$ if a graph isomorphic to $H$ can be obtained from a vertex induced subgraph of $G$ by contracting edges, discarding any loops or multiple edges that may form. There are many interesting families of graphs which are closed in the $C$-minor order but not in the minor order. Some examples are:

(1) The graphs which have independence number at most $k$;
(2) Interval graphs;
(3) Circular arc graphs;

(4) Graphs representable by intersections of line segments in the plane.

Several of these families (though not all) have characterizations in terms of a finite number of forbidden $C$-minors.

Does this partial order have polynomial time order tests, or is there some graph $H$ for which the problem of determining, for input $G$, whether $G$ has $H$ as a $C$-minor is $NP$-complete?

This problem was raised by Mike Fellows at the conference; here are some other comments and problems associated with $C$-minors.

*1:* The *Hadwiger number* of a graph $G$ is $h(G) = max\{n|K_n \text{ is a } C\text{-minor of } G\}$.

Because $K_n$ is a $C$-minor of $G$ if and only if $K_n$ is a minor of $G$, there exists a polynomial algorithm to determine, for fixed $k$, if $h(G) \geq k$. In fact, there is a practical $O(n^3)$ algorithm to accomplish this.

*2:* As shown in [1], Kuratowski's forbidden minor $K_{3,3}$ can be replaced by a few $C$-minors in the criterion for planarity. That is, there is a finite number of forbidden $C$-minors that characterizes planarity. Is there a criterion for selection of edges whose contraction would preserve these forbidden $C$-minors in a graph?

*Conjecture:* Contraction of an edge of a triangle incident to a vertex of least degree does not affect planarity.

*3:* The problem of whether a graph $G$ contains a clique of size $k$ is $NP$-complete for non-planar graphs. Is it possible to first find a contraction to a Hadwiger graph and only then to look for a clique which would be a $C$-minor (or subgraph) of the Hadwiger graph?

1. L. Basenspiler, *A note on forbidden minors*, Proceedings of the $250^{th}$ Anniversary Conference on Graph Theory, Indiana University-Purdue University, March 1986 (to appear).